Designed Technologies for Healthy Aging

Claudia B. Rébola

SYNTHESIS LECTURES ON ASSISTIVE, REHABILITATIVE, AND HEALTH-PRESERVING TECHNOLOGIES

EDITOR

Ron Baecker, *University of Toronto*

Advances in medicine allow us to live longer, despite the assaults on our bodies from war, environmental damage, and natural disasters. The result is that many of us survive for years or decades with increasing difficulties in tasks such as seeing, hearing, moving, planning, remembering, and communicating.

This series provides current state-of-the-art overviews of key topics in the burgeoning field of assistive technologies. We take a broad view of this field, giving attention not only to prosthetics that compensate for impaired capabilities, but to methods for rehabilitating or restoring function, as well as protective interventions that enable individuals to be healthy for longer periods of time throughout the lifespan. Our emphasis is in the role of information and communications technologies in prosthetics, rehabilitation, and disease prevention.

Designed Technologies for Healthy Aging
Claudia B. Rébola

ISBN: 978-3-031-00470-4   print
ISBN: 978-3-031-01598-4   ebook

DOI 10.1007/978-3-031-01598-4

A Publication in the Springer series
SYNTHESIS LECTURES ON ASSISTIVE, REHABILITATIVE,
AND HEALTH-PRESERVING
TECHNOLOGIES #6
Series Editor: Ronald M. Baecker, University of Toronto

Series ISSN 2162-7258 Print 2162-7266 Electronic

Designed Technologies for Healthy Aging

Claudia B. Rébola

SYNTHESIS LECTURES ON ASSISTIVE, REHABILITATIVE, AND HEALTH-PRESERVING TECHNOLOGIES

Ron Baecker, *Series Editor*

Abstract

Designed Technologies for Healthy Aging identifies and presents a variety of contemporary technologies to support older adults' abilities to perform everyday activities. Efforts of industry, laboratories, and learning institutions are documented under four major categories: social connections, independent self care, healthy home and active lifestyle.

The book contains well-documented and illustrative recent examples of designed technologies—ranging from wearable devices, to mobile applications, to assistive robots—on the broad areas of design and computation, including industrial design, interaction design, graphic design, human-computer interaction, software engineering, and artificial intelligence.

Keywords

aging, older adults, design, industrial design, interaction design, graphic design, human-computer interaction, software engineering, artificial intelligence, technologies, sensors, robotics, wearable computing, mobile computing, embedded computing, tangible computing, health, social disconnectedness, isolation, independent living, caregiver, smart homes, retirement communities

DEDICATED TO MY FAMILY IN ARGENTINA,
MY HUSBAND, AND MY AMAZING DAUGHTER

Contents

Acknowledgements

This book has been in the making since I started my investigations on the topic of design for aging. I am deeply grateful to two individuals who helped me initiate and develop my focus area. Elizabeth (Beth) Mynatt, then Director of the GVU Center at the Georgia Institute of Technology, was instrumental in my career through the center's seed grant program, which funded my initial exposure on designing technologies for older adults. Perhaps more importantly, the seed grant gave me the courage to follow my ideas, out of the normal. Moreover, I could not have been so successful without my colleague and mentor, Jon Sanford. Over the years, I profited from his guidance to elevate my design skills within a scientific approach.

This book became a reality when I received a faculty grant from the College of Architecture at the Georgia Institute of Technology. In particular, I wish to thank Dean Steve French, then Associate Dean for Research, for funding projects of this nature. I am grateful to the students who helped in this endeavor, Yoni Kaplan, who contacted contributors and performed the initial editing of submissions, and Luke Mastrangelo, who assisted with the graphic design layout for this book.

Special thanks to Ron Baecker and Morgan & Claypool Publishers for believing in this project. This book would not exist without their support. Most importantly, this book would not exist without the contributors. I owe particular thanks to them—their willingness, patience and intriguing work they in this needed area of design for our older adult population. I am also indebted to my friends, colleagues and students who suppprted me throughout the process.

INTRODUCTION

Introduction

DR. CLAUDIA B. REBOLA

This book is about cataloging designed technologies that have been designed for the healthy aging of our older adults. This project started as a simple literature review on what industry, researchers and entrepreneurs have done in the field of aging in relationship to technologies[1][2]. Soon it became apparent that there was a contradictory dilemma: it was difficult to find up-to-date information on the topic yet it was a timely and relevant topic for the growing population. Even though nowadays we are just clicks

---

1. Pirkl, J. J. (1994). Transgenerational Design: Products for an Aging Population. New York: Van Nostrand Reinhold.
2. Brawley, E. C. (2006). Design Innovations for Aging and Alzheimer's: Creating Caring Environments. Hoboken, NJ: J. Wiley.

away from having instant access to a plethora of information, it was challenging to find exemplary products on the topic of designed technologies for older adults, an aging population that is dramatically growing and eager to stay healthy and independent. The purpose of writing this book was then clear. There was a need to develop a catalog that can help a large audience identify work on the topic. This motivation was then converted into a research project.

The presented work is a result of a research study aimed at documenting designed technologies for healthy aging. A self-design survey was developed and delivered to participants using a snowball sampling technique. The survey contained questions including product name, authors, date of execution/release to market if applicable, product description, specifications, significance of the project, additional resources, and recommendations of designed technologies for aging considered exemplar. The goal was to reveal, in a more apparent manner, the current status of designed technologies that can serve as a starting point for discussion on trends in the field.

Products documented in the book met the following requirements: they must have been designed within the past five years, make use of digital interactive technologies to support activities[3], addressing older adult's activities of daily living (ADLs), instrumental activities of daily living (IADLs), and/or psychological and social functioning. Older adults have different needs that can be as basic as eating. Even a simple activity such as eating can be a major challenging task. Older adults also have necessary needs, including taking medications, mobilizing, caring for the home, to mention just a few. But there are also social needs[4], including talking to family members, participating in communities, and learning new skills. Designing products to meet these needs is a key factor toward addressing successful and healthy aging. We are currently in a technological era in which we are capable of turning imaginative ideas into reality. The book presents fascinating explorations of technology designs to support older adult's everyday needs. It documents the efforts not only of laboratories and businesses but also learning institutions and course outcomes toward designing technologies that

---

3. *Katz for the Association of Rheumatology Health Professionals Outcomes Measures Task Force, P. P. (2003), Measures of Adult General Functional Status: The Barthel Index, Katz Index of Activities of Daily Living, Health Assessment Questionnaire (HAQ), MACTAR Patient Preference Disability Questionnaire, and Modified Health Assessment Questionnaire (MHAQ). Arthritis & Rheumatism, 49: S15–S27.*
4. *Quadagno, J. (2007). Aging and The Life Course: An Introduction to Social Gerontology, McGraw-Hill Higher Education.*

enable healthy aging and facilitate social, communication, assistive, home and leisure activities. It contains well-documented and illustrative recent examples of designed technologies ranging from wearable devices, mobile applications to assistive robots on the broad areas of design and computation including industrial design, interaction design, graphic design, human-computer interaction, software engineering, and artificial intelligence.

There are four sections in this book to help navigate examples of designed technologies for older adults: *social connections designed technologies, independent self-care designed technologies, healthy home designed technologies* and *active lifestyle designed technologies*. Each designed technology is presented in four facing pages documenting the goals, technologies and studies conducted.

There are also references on the product in the form of website links and published articles. In addition, each technology is annotated with three acronyms to represent the type of design: *design in the marketplace* (M), *concept design* (C) and *prototype design* (P). Lastly, each product design is coded describing the main features of the technology use, including: *adaptive interfaces, broadband connections, data management, design research, embedded computing, graphic interface design, interaction design, internet connected, mobile applications, robotics, sensing technologies, social media, tangible interfaces, touch screen interfaces, universal design, vision interfaces, video communications, wearable devices, web applications, and wireless technologies.*

The products portrayed in each category are not mutually exclusive. There are certainly overlaps, sharing a committed interest of developing products within a user-centered approach with respect of making technologies usable and augmenting older adult's capabilities in the built environment.

M   DESIGN IN THE MARKETLACE

C   CONCEPT DESIGN

P   PROTOTYPE DESIGN

INTRODUCTION

Sections

Social

Connections

Research has shown that social disconnectedness is a major factor of depression, characterized by a lack of contact with others and indicated by situational factors, including small social networks, infrequent social interactions and lack of participation in social activities and groups.[1] To date, aging Americans spend approximately a third of their time sleeping, a third of their time in required activities, such as grooming, and the reminder of their awake time in leisure activities, in which approximately half of that time is spent watching TV and only 10% socializing.[2] Lack of interaction with others and social disconnectedness not only affect quality of life but also have negative effects on health that can lead to higher mortality rates among older adults.[3]

This issue poses the need to develop technologies that can address social disconnectedness in the aging population. The following examples from industry, laboratories and learning institutions are promising solutions to allow older adults to connect with others, create networks and participate in intergenerational and group activities.

"Good Night Lamp" is an unobtrusive interactive light in the shape of a house in which older adults can send messages to family members through tapping the light. Similarly, "InTouch" brings families together through a picture frame device. The uniqueness of these products is how the interactivity is built with familiar objects. "IC" also proposes solutions to address the aesthetical appearance of products, especially "Objects of Another Age" by designing affordances and making digital devices look more like what they do. "Jive" and "Communi-Table" bring a tangible, interactive approach toward solving problems of usability and digital interfaces. "CommCUBE" explores the opportunity of linking physical behaviors, such as writing, enhanced with technological

opportunities. "Clarity Ensemble" presents state-of-the-art technology in the design of an amplified telephone, that among several usability functions, features caption-enabled realtime transcription screens.

There are also projects that propose innovative approaches to technologies for socialization. "Storymaker, Storyteller" and "Hamefarers' Kist" propose technological solutions to bring generations together. Lastly, "Social Sewing" is a set of interactive products that connect older adults through craft activities.

Overall, this section showcases solutions geared to making technologies more usable for older adults. But more importantly, it portrays designs aimed at aiding older adults to better communicate with others and be more social by placing emphasis on the tangible and aesthetical aspects of technology applications.

1. Cornwell, E. Y., & Waite, L. J. (2009). Social Disconnectedness, Perceived Isolation, and Health Among Older Adults. Journal of Health and Social Behavior, 50(1), 31–48.
2. Federal Interagency Forum on Aging-Related Statistics. (2012). Older Americans 2012: Key Indicators of Well-Being. Washington, DC, USA.
3. National Institute of Mental Health. (2007). Older Adults: Depression and Suicide Facts (Fact Sheet). From http://www.nimh.nih.gov/about/connect-with-nimh/index.shtml.

M

Good Night

Lamp

ALEXANDRA DESCHAMPS-SONSINO   *2012*

The "Good Night Lamp" is a family of Internet-connected lamps. The system is comprised of a single Big Lamp and a series of Little Lamps, which are connected to a wireless home network. Helping families connect to an older adult living alone is a challenge. The Good Night Lamp is a light in the shape of a house that helps families stay connected regardless of their location. Older adults can use the lamps to communicate with family members.

By switching a Big Lamp on through the chimney, a network of Little Lamps will light up. Giving the Little Lamps to family or friends lets them know when the Big Lamp is turned on.

With a simple, unobtrusive, and friendly interface, older adults can send messages through the light, inferring meaning with an everyday gesture, to their families. There are multiple applications of the Good Night Lamp and its simple point of presence that can be socially constructed.

The light message can signal readiness for an action such as a Skype call, going to bed, safe at home, or that they are simply reading before going to bed, but their family can see that they're around, they are safe, and they are active. With certainty, older adults can feel part of a community, knowing they are not alone in their home environment.

MORE INFORMATION:
http://goodnightlamp.com
http://makezine.com/2013/04/04/alexandra-deschamps-sonsino-making-the-good-night-lamp/

Content in this section provided by Alexandra Deschamps-Sonsino.

Data Managment, Graphic Interface Design, Internet Connected, Touch Screen Interfaces, Wireless Technologies

P

InTouch

RONALD BAECKER   *2010–2014*

The "InTouch" is a simplified digital picture frame with communication capabilities accessed through photographs of loved ones and friends. It is designed to encourage communication and interaction of older adults at risk of social isolation. Older adults use the Internet for a variety of purposes: to access information, for entertainment, and for keeping in touch through email, instant messaging, and desktop video conferencing. Yet, there is a large number of older adults

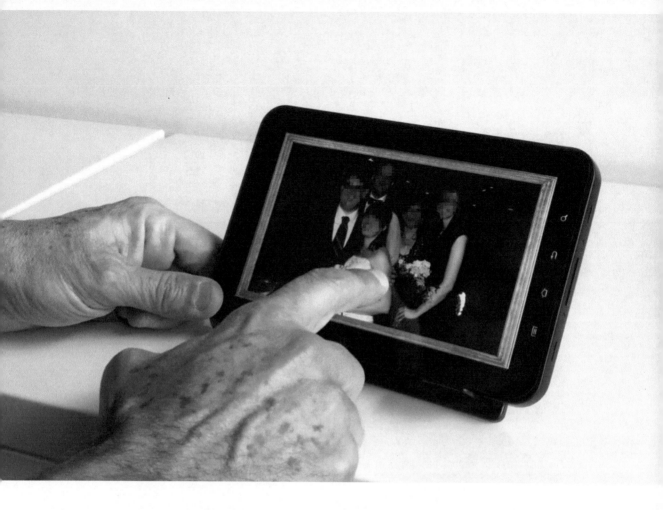

who do not make use of such communications technology and are isolated. InTouch is designed to address the learnability and usability of communication technologies, accommodating the needs of computer-shy older adults. InTouch features a familiar and simple design via a picture frame that can be used in different environments such as the older adult's home as well as hospital rooms or other locations where computers are not readily available or easily usable. The frame is a small tablet computer running a customized application, displaying family and friends in the frame. When pictures are touched, the frame sends an email message to one or more loved ones with short messages such as "I'm thinking of you." Recipients are then prompted to respond by

1

Frame is touched by an isolated person

2

"I'm Thinking of you"

Message is transmitted to family members in picture frame: *"I'm Thinking of you"*

3

Message is received by family members through email

4

Video message is recorded by a family member

5

Message is sent back to picture frame

6

Frame owner is notified of a new message, which is then played back

P

sending a video, a picture, a voice message, or a text message using their computers or mobile phones. After the content is received, the picture frame indicates the new message, both visually and audibly. InTouch supports asynchronous communication and is easily accessible, having a simple, intuitive and robust design. It features customizable reciprocity, supporting different types of messages. The InTouch owner can easily send a wave {"I'm thinking of you!"), a voice message, a photo, or a video. No typing is required.

The original idea of this device stems from observing the increasing isolation of a family member in the hospital whose body was shrinking due to MS, through a consideration of the needs of older adults suffering from dementia, and from communication studies of people living at home with chronic pain. In all cases, the affliction caused barriers to communication thereby affecting the quality and quantity of contact with loved ones.

Cellular Mobile Device

Video Message

I'm thinking of you

Server

Message from frame
"I'm thinking of you"

Digital Picture Frame

Wireless (TCP-IP)

Video Message

Personal Computer

InTouch is also a convenient tool for the owner's loved ones and aides to support families managing busy lifestyles and caregiving responsibilities, especially when geographically distant.

MORE INFORMATION:

Work on 3 prototypes called Families In Touch, Ringo, and InTouch also included contributions from Kate Sellen, Steve Tsourounis, Anselina Chia, Ian Stewart-Binks, Marc Fiume, Spencer Beacock, Chris Arnold, and Garry Beirne, and was made possible by grant support from NSERC, Google Research, the GRAND NCE, Revera Inc., OCE, and CC.

Technologies for Aging Gracefully Lab (TAGlab), University of Toronto: www.taglab.ca

Baecker, R.M., Neves, B., Sellen, K., Crosskey, S., & Boscart, V. (2014). Technology to Reduce Social Isolation and Loneliness. Paper presented at the ACM ASSETS'14: The 16th International ACM SIGACCESS Conference on Computers and Accessibility. Rochester, NY, USA.

Benjamin, A., Birnholtz, J., Baecker, R., Gromala, D., & Furlan, A. (2012). Impression Management Work: How Seniors with Chronic Pain Address Disruptions in their Interactions. Proceedings from ACM CSCW'12: Conference on Conference on Computer Supported Cooperative Work. Seattle, WA, USA.

Content in this section provided by Ronald Baecker.

C

Objects of

Another Age

EVA RIELLAND   2011

The "Objects of Another Age" are a series of products making use of tangible interfaces for common communication tools found in modern computers and tablets. Each object represents a different task and has a visual and cultural cue to perform the task.

Write, view, upload images, and print are the tasks that can be performed by these objects. These objects include a frame used for video chat, a mailbox to exchange e-mails, a printer and a visual stand that presents an image received by e-mail.

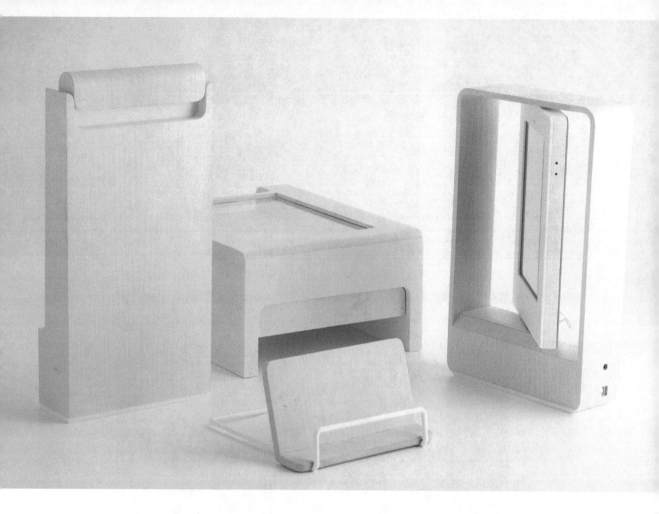

The use of each object utilizes a gestural interaction that minimizes the use of graphical user interfaces commonly found in digital devices. An intuitive gesture for each object builds on past experiences in order to help users understand functions through tangible and visual interactions.

Objects from another age aim to rethink high tech tools and make them easily understood by all, and particularly by older adults. The products were designed to reduce learning curves an maximize the use of technology devices within the older adult population.

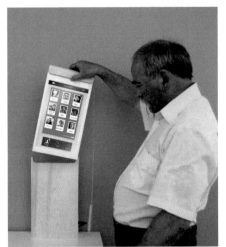

C

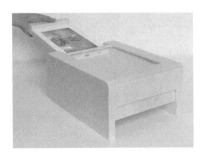

The design of the system is based on observations of real world habits and needs to demonstrate that the consideration of aging may be a source of innovation and a positive value for creating objects.

MORE INFORMATION:
Project funded by VIA and exhibited at VIA (2010) and "Paris Design Week" (2011).
www.evarielland.com

Content in this section provided by Eva Rielland.

Graphic Interface Design, Internet Connected, Mobile Applications, Touch Screen Interfaces, Wireless Technologies

P

IC

MIKAEL JOHANSEN 2010

The "IC" is a handheld videophone tablet device for older adults. The device presents digital content from different media, such email, in a more friendly interface for older adults. The content is simplified and presented in a format that reduces cognitive overload when operating software interfaces and tablet devices. Messages can be handwritten or voice recorded. In addition, the system automatically sends messages to the receiving channel.

IC allows older adults to use communication applications and tablet devices in order to stay better connected with family members located in different regions of the world.

IC helps older adults enter the arena of social media, allowing them to be more active and independent with digital content. The IC device aims to help older adults stay in touch regardless of age.

P

The tablet device is a bidirectional touchscreen that runs in the Android operating system. By using a bidirectional screen as a web- camera and a touch screen this product gives the experience of actual eye contact.

The device system operates through a 4G network, in which no installation is needed. In addition, during first time use, the device automatically creates email and social media accounts from a suggested list.

SOCIAL CONNECTIONS

The design makes use of environmentally friendly wood finishes for a softer look compared to traditional technology products. The goal is to facilitate a better acceptance of technology products by older adults.

MORE INFORMATION:
The product has been recognized by Yanko Design and F5 Trend book. It has also been exhibited in the Shanghai Expo 2010.

www.mikaeljohansen.com http://www.yankodesign.com/2010/06/25/making-it-easy-to-communicate-with-grandma/ http:// www.trendhunter.com/trends/ic-social-comminucation-tool

Content in this section provided by Mikael Johansen.

Graphic Interface Design, Internet Connected, Social Media, Tangible Interfaces, Touch Screen Interfaces

P

Jive

BEN ARENT 2008

The "Jive" is a communication device designed for older adults. Jive is a tangible, interactive display that aims to solve issues that keep older adults from accessing online content and communication tools. Connecting to the web, creating and sharing contacts, and mouse navigation all hinder older adults from actively using web tools.

Due to decreased dexterity and vision, everyday computers and web tools may not be user friendly to older adults. Lack of access to

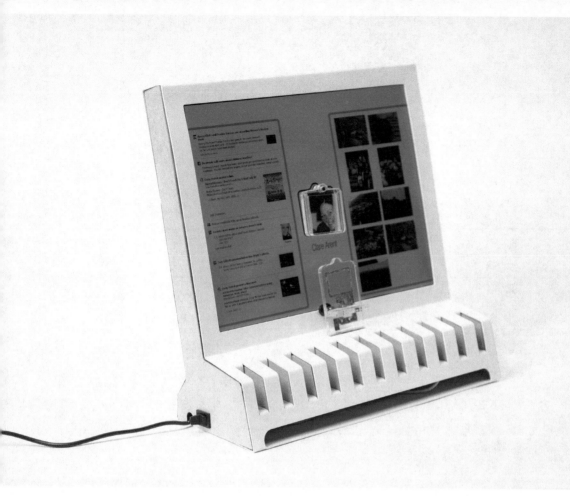

web tools may lead to a decrease in social interactions since family and friends spend much of their social life online. Jive makes these interactions more accessible by making the experience simplistic and tangible. Jive allows the user to access the web easily without any setup. This is accomplished by pre-configuring the device with an internet service provider, enabling a plug and play experience.

SOCIAL CONNECTIONS

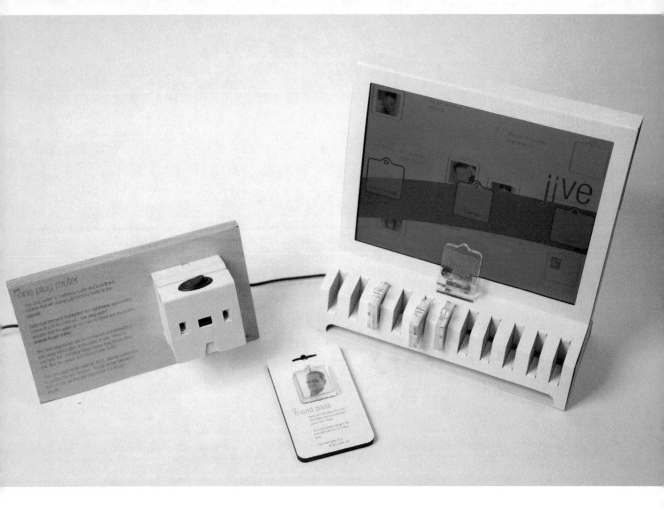

To create contacts, the user's friend will associate their online content with a tangible tag. The tag serves as the primary interaction for the older adult.

If users wish to view the information of a particular contact, they simply place the tag in physical slots on the display. This eliminates the need for mouse navigation and will only display one contact at a time to avoid complications.

MORE INFORMATION:
http://www.benarent.co.uk/

Content in this section provided by Ben Arent.

M

Clarity

BIG BANG INC. FOR CLARITY, PLANTRONICS   2010

The "Clarity® Ensemble™" is an amplified telephone that displays a transcript of a phone conversation in real-time on a large 7-inch color touchscreen. The amplified handset provides loud and clear sound of up to 50 decibels using the same state-of-the-art technology found in high-end hearing aids. Users can have the settings and contact numbers of the phone customized automatically using the built- in "ClarityLogic" customer support system.

The phone features comprehensive accessibility features for older adults. It utilizes "ClearCaptions" phone captioning service. It also features amplification up to 50 dB, making it ideal for people with severe hearing loss. The handset is light and very easy to hold with a small diameter barrel. The ear cup is specifically shaped to accommodate hearing aids.

The large 7" touchscreen LCD display provides bright, vivid images and large captions for people with low vision. The adjustable stand can accommodate multiple angles to avoid glare and ergonomic issues. The product interface is simple to use, easy to navigate, adjustable (button, number and font size) and features high contrast graphics.

M

Creating the product was a balance, utilizing expertise in Industrial Design, User Interface Design, and Mechanical Engineering. A critical design goal was to identify an appearance language that communicated sophisticated technology without alienating the target user who tends who might be uncomfortable with the changes. The simple, compact design focuses on the display and handset.

While it looks like a traditional telephone, it blends computing and tablet features in the telephones. The design keeps telephony cues, like the handset and speaker very pronounced, while incorporating a touch panel display in a familiar telephone angled shape.

Menu Screen

Quick access to menu features from home screen.

Incoming Call Screen

The red border flashes to help alert the user to an incoming call.

Example of an incoming call in progress with the volume controls.

Developing a simple and intuitive user interface experience for non-touch and non-computer users was imperative. Simple and self explanatory icons with high contrast colors were used while keeping screens uncluttered.

Special attention was given to customization of the interface. Users can easily adjust text and number size and color-coordinated sections including phonebook, messages, outgoing/incoming calls and settings.

MORE INFORMATION:
http://bigbangip.com/
http://clarityproducts.com/

Content in this section provided by Steve Meinster.

Graphic Interface Design, Interaction Design, Internet Connected, Social Media, Touch Screen Interfaces, Wireless Technologies

C

CommCUBE

EMILY KEEN AND CARRIE SMITH   2011

The "CommCUBE" is a universal and multi-faced communication device that is used as an environmental product in the home. The device is designed for the older adult persona to facilitate communication interactions with friends and family members.

CommCUBE also allows the user to manage task activities through an easy to use and friendly calendar. Users make use of touch and tangible interfaces for specific functions including contacts, photos, mail, and schedule. Each interface interacts

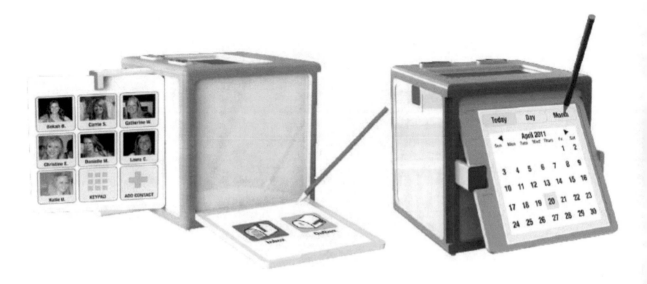

with one another as it syncs to any wireless communication device that it placed on the top surface of the CommCUBE. As such, this product allows the user to input data from his/her wireless device into the system.

CommCUBE is designed to enable older adults to transition comfortably and easily to adoption of current technology devices via touchscreen interfaces. The icons and text are simplified and designed in order for the users to quickly navigate through the interfaces and to enable easy-reading of the content.

C

Older adults with reduced vision and fine motor control may find an issue with the simplest of interactions on a small wireless communication device.

The CommCUBE allows them to view all the information on their cellular phone, without navigating menus or worrying about small buttons.

SOCIAL CONNECTIONS

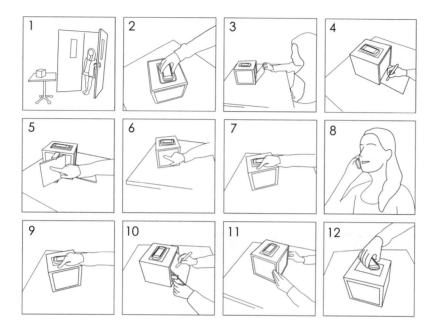

1   User gets home, message is waiting

2   User places phone and syncs the device

3   User answers message and receives photo

4   User tags photos

5   User opens "Contacts" screen

6   User finds and selects contacts

7   User calls grandchild

8   Grandchild reminds user of an appointment

9   User ends call

10   User enters appointment into schedule

11   User closes product

12   User retrieves phone

MORE INFORMATION:
http://www.id.gatech.edu/node/5655/lightbox2

Co-author includes Alison Pak.

Content in this section provided by Emily Keen.

C

Communi

HSIEN-HUI TANG 2010

The "Communi-Table" is a desktop lamp that allows older adults to create digital content from analog interfaces. Communi-Table explores human-computer interaction opportunities based on the desktop metaphor. It turns pen and paper interfaces in the physical environment into a digital interface. Communi-Table aims to bridge the gaps between older adults with technology illiteracy and other generations with advanced digital literacy.

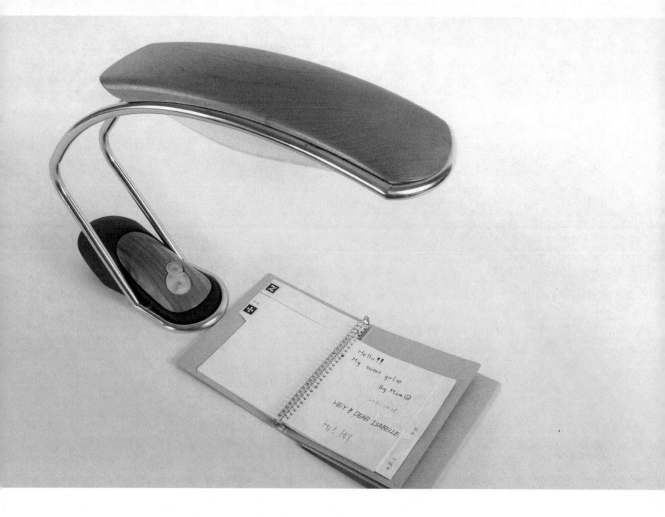

Older adults are more familiar with pen-and-paper, memos, and physical calendars. Younger generations are accustomed to computers, e-mail clients, and web-calendars. This divide is becoming more acute as technology advances, hindering communication interactions among generations.

By maintaining an analog interface, this ensures that older adults continue with their modalities while interacting with others in the digital realm.

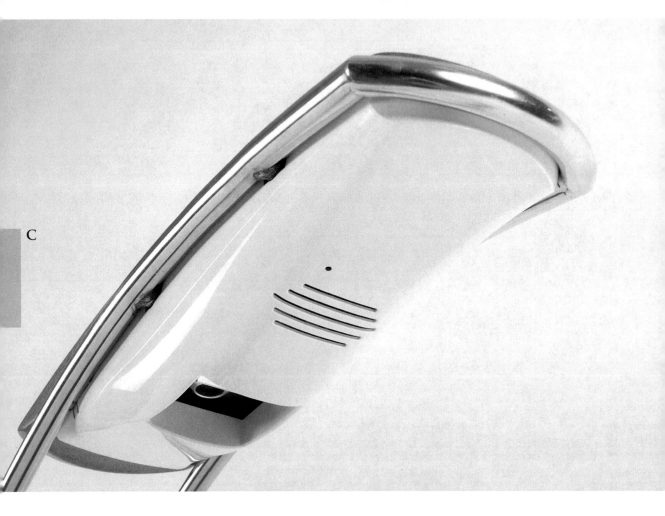

C

Communi-Table is a computer-embedded device taking the form of a desktop lamp. The base of the lamp houses the computer while the lamp head houses the projector and camera. The embedded camera captures the images of physical photos and words as the projector displays digital images and words on the table.

Communi-Table also utilizes a notebook, which has different separator pages with augmented reality tags. These tags are barcode-like that only the camera can see. When each tagged page is read, a new interaction is activated.

Communi-Table has three main functions for communication, being e-mail, photo and blog.

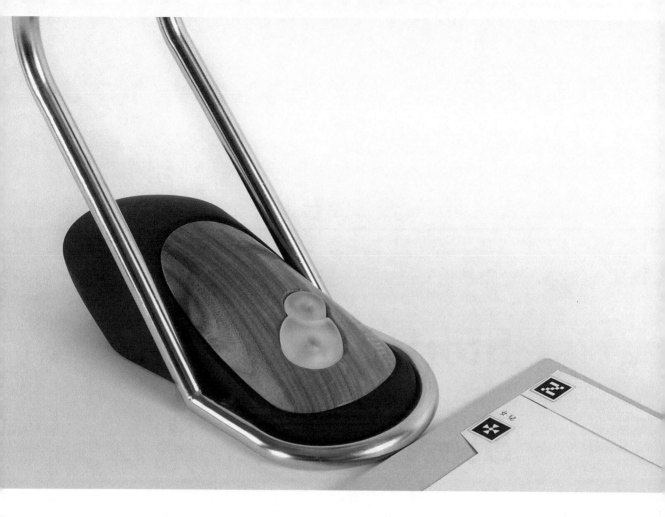

When the user flips to a page related to e-mail, the camera will transfer the contents of the notebook page into an e-mail. Similarly, photo albums or blogs will be activated by accessing the associated pages. To confirm the content, the user will press the button located on the base of the lamp.

E-mail, photo and blog functions are essential for the aging population who needs to communicate with their relatives in the digital realm. With a seamless input and output, Communi-Table can play a significant role assisting communication between veteran technology users with new digital and internet-based devices.

MORE INFORMATION:
Project developed at the National University of Taiwan of Science and Technology, National Taiwan University and Chang Gung University. Credits to: Mu-Chern Fong, Chin-Chun Kao, Rong-Hao Liang, Ying-Ling Chen, Cheng-Wei Chen, Yi-Hsin Cheng, Gwen Hsiao, Yang-Bee Lee, Chih-Ying Yang, Wen- Chieh Fang, Poming Chen, Ching-Yi Chan, Guan-Hong Chen, Jyong-Cheng Liao, Ting-An Lai, Ya-Chih Li, Yu-Fang Tai, and Shu-Min Chang.

http://ditldesign.sqsp.com/design-projects/#/design/2010-npui/

Content in this section provided by Hsien-Hui Tang.

P

Storymaker,

Storyteller

Neil Dawson, Natalie Montgomery,
Lee Murray and Joanna Montgomery   2008

The "Storymaker, Storyteller" are a pair of interactive objects designed to create a connection between individuals from two generations, such as a grandparent and their grandchild, through stories and pictures. Inspired by the design of 1970s electronic appliances, these devices allow a young person distant from their grandparent to enjoy the experience of a spoken story, accompanied by pictures shown as they would be on a traditional mechanical slide projector.

The Storymaker is a handheld device in which the grandparent can place their slides. Reminiscent of a traditional slide viewer, its familiarity and simple mechanics allow an older adult to operate it with ease. one press of a button, the slides and stories spoken by the grandparent begin to be captured. As the recording goes on, the inserted slide can be replaced to add more pictures to the story.

When the button is pressed again, the pictures and audio are transmitted to the Storyteller as a single "presentation."

The Storyteller is a desktop projector which receives and archives the presentations. With a simple rotary control, grandchildren can scan through the stories, and a single buttonpress begins the slideshow.

When the presentation is playing, the granparent's voice fills the room, and the accompanying photographs are projected at the same moments they were inserted into the Storymaker.

It was important for the project to preserve the sensations that come with being in a darkened room, listening to the grandparent talk. For this reason, there are no pauses, fast forward or rewind controls—pauses and mistakes are part of the charm.

Storymaker, Storyteller
revives these disappearing
experiences, and allows sharing
of precious stories that might
otherwise be lost. A grandparent
can speak in the technology of
their day, and a grandchild can
listen in the technology of theirs.

MORE INFORMATION:
Project leader Graham Pullin, Duncan of Jordanstone College of Art & Design, University of Dundee.

Pullin, G., Rogers, J., Banks, R., Regan, T., Napier, A., & Duplock, P. (2011). Social Digital Objects for Grandparents. Proceedings from Include'11: Conference on Inclusive and People-Centred Design. Royal College of Art.

Content in this section provided by Graham Pullin and Neil Dawson.

Interaction Design, Internet Connected, Sensing Technologies, Touch Screen Interfaces, Tangible Interfaces, Wireless Technologies

P

Hamefarers'

Kist

Hazel White and Paul MacKinnon   2009/10

The "Hamefarers' Kist" is a working prototype developed following research in Shetland—a group of islands that lie halfway between Scotland and Norway famous for its distinctively patterned Fair Isle knitting. Hamefarer is the local term for a Shetlander who returns to visit the islands. Many young people leave Shetland to study or work and don't return—creating a diaspora of tech-savvy younger people and leaving behind older, less digitally adept relatives. The Kist is designed to enable the younger population to

connect and share events with their older Shetland relatives. It demonstrates how geographically dispersed family members and friends can share online content using traditionally knitted objects. Images and messages uploaded from anywhere using a smartphone or tablet will be received by the kist and viewed by older adults. The knitted form of the kist provides a familiar object through which the content is viewed. This concept can also help other groups (i.e. healthcare professionals) to re-imagine methods for sharing information and services to older adults with complex communication needs. A small decorative box contains a series of "knitted remotes" that are uniquely patterned knitted "pincushions" containing RFID tags.

P

Placing one of these pincushions near the RFID reader in the box activates an application that calls up online content and displays it on the micro display in the box lid. The user can then browse through content using the capacitive screen.

Until 2010, all Shetland schoolchildren learned to knit traditional Shetland Fair Isle patterns and, therefore, have pattern recognition skills which remain with them for life. The familiarity and resonance of the complex knitting patterns used in the Hamefarers' Kist enable older adults to easily create associations between the knitted pincushions with people or events.
The tactile pincushions call up related online content eliminating the burden of remembering websites or how to log on to a computer. The content could be generated and shared by

grandchildren located in a different geographical location.

Images, video or messages are uploaded to the cloud and appear in their grandparent's kist. The older adult then remembers which pincushion is associated with each relative or friend and knows to change out the kist in the box to see new content for each person. Both the Hamefarers' Kist and Kist project are universally designed product service systems that can be accessed by everyone from older adults, people without literacy skills and very young children.

Kist project is a future development of the Hamefarers' Kist. Families and caregivers at the Children's Hospice Association of Scotland are creating tactile tagged objects which share the lives and loves of children with complex communication needs.

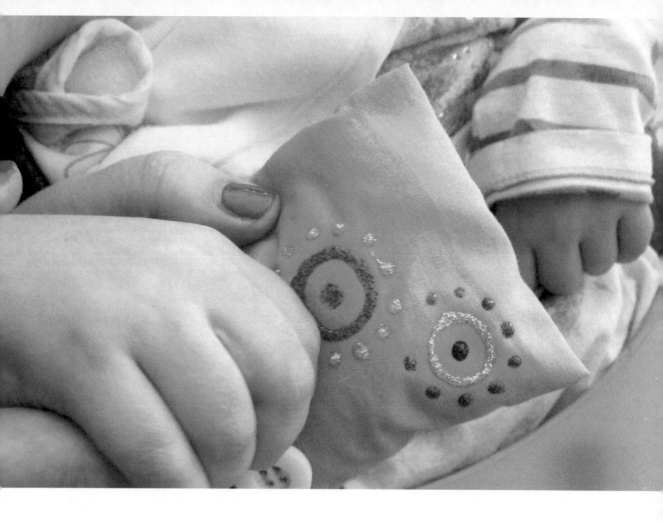

MORE INFORMATION:
http://hazelsnotes.wordpress.com/interactive-craft/knitted-remotes/
http://elusivesprite.squarespace.com/phd_journal/2010/4/9/hamefarers-kist-knitted-remotes.html

Content in this section provided by Hazel White.

Embedded Computing, Interaction Design, Tangible Interfaces, Universal Design

P

Social

Sewing

MICHAIL VANIS   2009

The "Social Sewing" is a system of interactive products resulted from the social digital objects for grandparents project collaboration between product design and interaction design students from the University of Dundee and Microsoft Research.

The project focused on designing products for older adults exercising inclusive design to catalyze radical thoughts of future roles for digital technology. Social Sewing was one of the results of the collaboration.

Social Sewing helps users and their circle of seamstresses to feel in touch during the activity of sewing. The concept aims at restoring the connection that seamstresses have when working in the same room.

Miniature sewing machines are used to represent the sewing machine of one of their friends.

When the friends' sewing machines turn, little motors in the miniature models rotate its wheels and move its needles up and down. The miniature sewing machines also plays a sewing sound mimicking the activity of sewing. The result is a chattering sound that is played in the background. The sounds are played through a loudspeaker located in a shelf.

Maria

Anna

Meropi

Despina

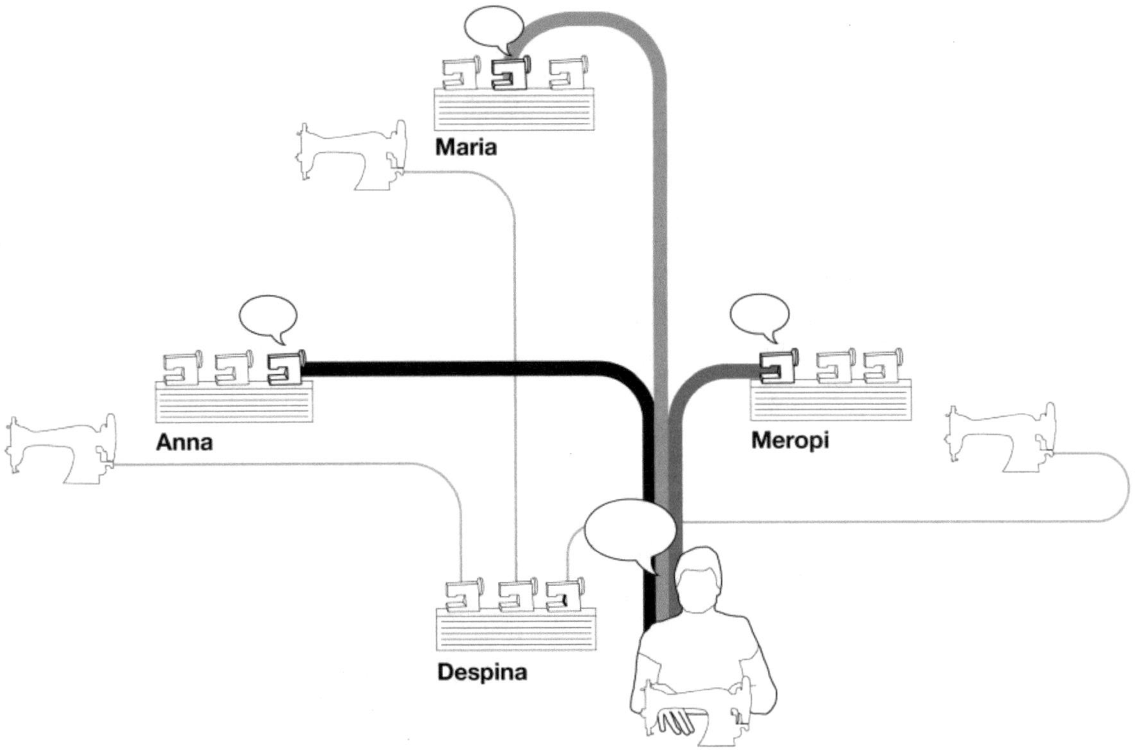

The concept uses dedicated Internet-enabled products. Arduino boards communicate wirelessly and are programmed to make the sewing machines behave the way they do.

There is a peripheral sense, in movement and sound, of the other women's sewing machines, which may just be companionable. But it could be also interpreted as an indication of how busy others are or the type of stitching (i.e., button holes) they are working on.

As each networked miniature
sewing machine mimics
the activity of the friends' own
sewing machines, it restores the
shared activity seamstresses used
to enjoy when they sewed together
in the same workshop.

MORE INFORMATION:

Project leader Graham Pullin, Duncan of Jordanstone College of Art & Design, University of Dundee. Credits to students Michael Vanis, Ruth Tullis, Anna Rendhal, Brian Matanda, Philip Gordon and Christopher McNicholl.

Pullin, G., Rogers, J., Banks, R., Regan, T., Napier, A., & Duplock, P. (2011). Social Digital Objects for Grandparents. Proceedings from Include'11: Conference on Inclusive and People-Centred Design. Royal College of Art.

Content in this section provided by Graham Pullin and listed paper.

Independent

Self Care

Independence is one of the most important aspects for a healthy aging. The Center for Technology and Aging[1] lists seven major application areas including medication optimization, remote patient monitoring, assistive technologies, remote training and supervision, disease management, cognitive fitness and assessment technologies and social networking. These areas respond to the older adult's needs such as: How to reach other people in an emergency? How to remember to take the right medications at the right time? How to remember to do basic activities such as eating? How to let families know yu are OK? Central to these questions is the older adult ability to *take care* of activities and situations. There is a need to generate innovative products to help older adults to live independently for longer by not taking care of them but allowing them to take self care with the help of technology applications and a support network. The following products represents examples of technology interventions to help older adults take care of their lives and health.

"ode" is an unobtrusive alert system designed to promote eating with older adults through enticing scents.

"GlowCap" proposes a practical approach toward medication management and alert systems. The product allows refills as easy as the push of a button and reminds older adults about their medications. Similarly, "Memo Pillbox" proposes practical solutions for medication adherence through a clever application of design principles such as color and shape.

"Genesis DM" is a remote patient monitoring product that tracks biometric information about the user and provide suggestions related to input data. "Genesis Touch" also offers the tracking of a user's information, but using tablet devices enhanced with video interactions between the user

and care provider in monitoring products, "Vital Signs Camera" uses breakthrough technology that allows older adults to measure their vital signs in an easy, unobtrusive and contactless way by using the camera from mobile devices. Besides remote monitoring and tracking tools, The "Clock Reader" presents a preventive tool that allows older adults to perform simple everyday exercises to learn about early signs of dementia.

Other products include "Talk" featuring a simplified version of wireless mobile devices. The product is an improved version of emergency buttons, but aimed at reducing mobile devices to its essentials. Lastly, unique products such as "Babyloid", a portable robot, remind us of the need to design tools to increase the feeling of purpose in the older adult's life by giving them a caregiving role.

Overall, this section showcases products designed to care for the self when living independently. Featured products include solutions aimed at remaining older adults to perform everyday simple tasks and facilitate communication with others in need of assistance.

1. Center for Technology and Aging (2009). Technologies to Help Older Adults Maintain Independence: Advancing Technology Adoption. From http://www.techandaging.org/briefingpaper.pdf.

M

ode

BEN DAVIES, LIZZIE OSTROM, MARK MORGAN AND SIMON LEVI   2012

The "ode" is a well being product that creates a link between the power of scent and the relationship with food. It helps older adults maintain an interest in eating, addressing the effects of malnutrition and promoting quality of life.

In the same way that the scent of freshly baked bread can stir feelings of hunger, ode releases high-quality bespoke food aromas. From bakery to fresh orange, scents migrate into living spaces to help stimulate appetite.

It is discreet compared to obtrusive alerts or monitoring systems. The product's food aromas act to create an environment in which people will want to eat. Many of the fragrances have been created to appeal to older adults, based on their favorite foods.

The device is small enough to be placed pervasively in the home. It can be placed on a bookcase, by a potted plant, or even next to the phone or TV, where it will quietly do its job.

M

There are three scents encased in a two part housing. Small containers are loaded after each cycle. This allows the user to test different scents and decide the time when they are activated.

To aid in building a regular appetite, ode is most effective when setup around mealtimes and daily routines.

Initial research shows positive results from 50% of people who were experiencing a decline in weight. Users reversed their trend showcasing a healthy weight increase once ode was introduced.

The product was designed to address the needs of older adults. ode was conceived by working closely with people living with dementia and care organizations

in order to understand the unique challenges of an aging population.

Malnutrition and weight loss is a high-risk condition for older adults impacting not only physical frailty but also agitation, irritability and aberrant motor behavior. ode addresses this issue as a gentle and enjoyable reminder for eating.

MORE INFORMATION:
http://www.myode.org
http://www.bbc.co.uk/news/health-17845782

Content in this section provided by Ben Davies.

M

GlowCap®

The "GlowCap" is a medical adherence solution for older adults. The product allows refills as easy as the push of a button and reminds older adults about their medications. GlowCap fits most prescription bottles and uses light and sound reminders to signal when it is time to take your medications. Inside the cap, a chip monitors when the pill bottle is opened and wirelessly relays alerts, through a mobile broadband network, to older adults or their caregivers.

Glowcap unique solution uses light and sound notifications. The lids glow and escalate from subtle to insistent. The devices glow, then make noise, then send a text notification or dial a home phone. There is also a social incentive. Each week the device sends an email to a friend or family member summarizing which days the person did or didn't take their medication, alongside educational content about their condition. Glowcap also features an accompanying reminder light that plugs easily into any outlet around the home. The device works in tandem with the lids by glowing orange at the time of the scheduled medication, and sends immediate updates to the user or caregiver.

A push button at the base of the lid makes refills easier. When the button is pressed, Glowcap lid sends refill requests to the local pharmacy. The system sets up an automatic call back to confirm the refill. Older adults simply need to pick up their medications at the pharmacy.

In 2009, Mass General Hospital recruited 200 people to use a Glowcap for a once-a-day heart disease medication. Data from this six-month randomized clinical trial showed a huge behavioral change caused by the GlowCap— an average adherence of over 95%!

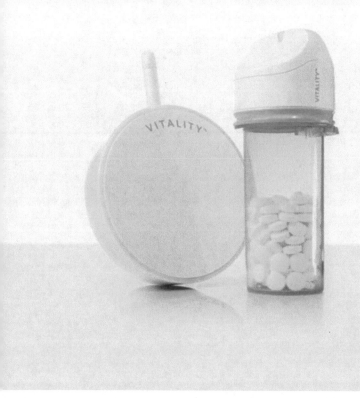

With automated reminders, push-of-a-button refills, personalized reports and real-time updates, Glowcap creates a full circle of care.

MORE INFORMATION:
http://www.vitality.net/pharma.html
http://www.glowcaps.com

Content in this section provided by the listed links and David Rose.

C

Memo

Pillbox

Joules Toulemonde   2012

The "Memo Pillbox" is an intuitive pill box, which is adjustable, adaptable and easy to carry. Memo is for older adults needing help to remember to take medicine. The product not only reminds older adults when to take but also how to take the medicine.

Due to its portability, Memo allows older adults to enjoy social, professional and sport activities outside their homes while reminded about their medicine needs.

Memo links the product to the
prescribing system. Doctors
have the ability to prescribe the
quantity, time and frequency
through a software system linked
to the product boxes.

PRESCRIBE / ORGANIZE   CARRY   ALERT / REMIND   TAKE THE RIGHT DOSE
AT THE RIGHT MOMENT

C

Each medicine is stored to a colored box. Each colored box is also associated with colored LED lights. The lighting system uses colors and blinks that are relaxing yet indicative.

Memo uses colored LEDs to remind and inform users about the time and quantity of medicine to take. The light indicates the color of the box, hence medicine that the user has to take.

This product was originally conceived to favor the insertion of older adults in society. Observations of older adults' behaviors, especially older adults with Alzheimer's disease led the design of the product.

MORE INFORMATION:
http://www.behance.net/gallery/PILL-BOX-FOR-ALZHEIMER-PATIENTS/3214869

Content in this section provided by Jules Toulemonde.

M

Genesis DM

HONEYWELL 2007

The "Genesis DM" is a remote patient monitoring tool that tracks biometric information about the user. Based on user input, the database can provide suggestions related to common symptoms. The device is also designed for use with peripheral monitoring tools such as digital scales, blood pressure cuffs, and pulse oximeters. The data input from these tools is sent to a clinical dashboard by the Genesis DM and added to the user's health profile.

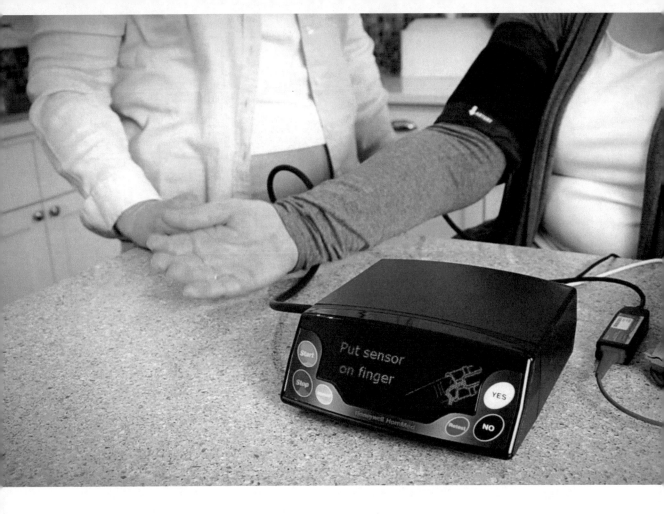

The information from the profile is viewable via the "LifeStream Manager", a clinical dashboard for care providers, giving them access to trends and unusual activity. The Genesis DM is designed around simple user interfaces. Six buttons with clearly defined functions are used to answer questions.

The user responds "yes" or "no" to questions, making it useful for older adults who are not accustomed to digital interfaces. LifeStream tracks the patient input and adjusts the questions and education provided accordingly. Questions and prompts can be also programmed to gather information for specific medical conditions.

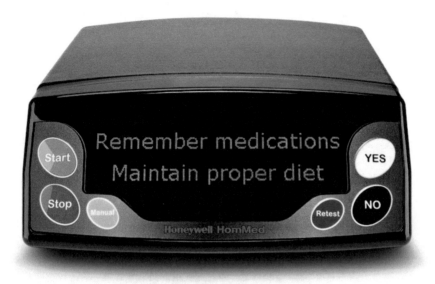

Visual and audio cues also make this product more accessible by offering multiple options for achieving the end result. The product offers male and female audio voices and supports over 15 languages. The language is presented on a 6th grade reading level to accommodate users.

The Genesis DM operates in combination with the Lifestream Management Suite, which provides encrypted databases that care providers can use to provide patient oversight. The Genesis DM uses landline connectivity to transmit information, but has the option of using a cellular network by adding a cellular modem to the product configuration.

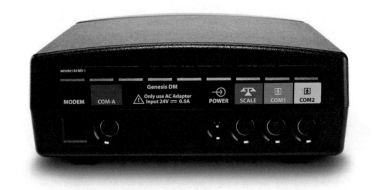

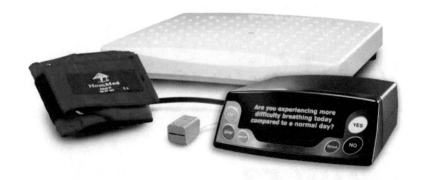

Research was done to develop the product form, which emulates an alarm clock as potential users did not want a device that resembled medical products. Weighing in at 2 pounds, the product is small enough to fit on a nightstand but has a low profile that cannot easily be knocked over.

MORE INFORMATION:

The product was designed together by Honeywell teams in India and Milwaukee, Wisconsin. Honeywell HomMed is the proud winner of a 2008, Medical Design Excellence Award.

http://www.hommed.com/lifestream-products/genesis-dm/

Content in this section provided by Stacey Force.

M

Vital Signs

Camera

VINCENT JEANNE AND MAARTEN BODLAENDER   2011

The "Vital Signs Camera" is a mobile application designed to perform heart and breathing rate measurements. The Vital Signs Camera uses breakthrough technology that allows older adults to measure their vital signs in an easy, unobtrusive and contactless way by using the camera from mobile devices. The application uses computer vision methods to identify key health information.

The camera is able to capture heart rate data just by analyzing the color of the user's face.

The camera detects small changes in the user's face color to measure heart rate with beat-to beat accuracy.

Likewise, the camera tracks the user's breathing rate. It can accurately track breathing rate by measuring the motion of the user's chest. The application can also measure two people at the same time.

The interface is unobtrusive and does not require the user to hold the tablet computer in order to accurately capture data. The technology can measure up to 15 m with a uEye camera, and approximately 1 m with a tablet camera.

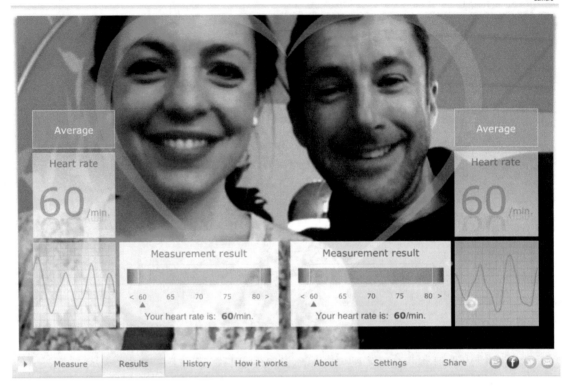

Information is stored for long term visualization of statistics and can be exported through itunes file management.

Integrating this product into devices that are already widely used makes the product adoptable. By minimizing the interactions required for effective use of the application, the product becomes more usable to all ages and abilities.

Accessibility features such as these
are embedded into the architecture
of the software as the product
does not require initial setup or
calibration.

MORE INFORMATION:
Product designed at Philips and Centagon.

www.vitalsignscamera.com
https://itunes.apple.com/us/app/vital-signs-camera-philips/id474433446?mt=8
http://innovation.philips.com/pressreleases/nurturing-bg_vital_signs_camera/index.html

Content in this section provided by Maarten Bodlaender.

M

Genesis

Touch

HONEYWELL   2011

The "Genesis Touch" is a mobile remote patient monitoring device that tracks biometric information about the user and offers a platform for "one touch" video interactions between the patient and care provider. Based on user input, the Genesis Touch can display education to the user related to common symptoms. The device is designed for use with bluetooth peripral monitoring tools such as digital scales, blood pressure cuffs and pulse oximeters.

The data gathered by the Genesis Touch is sent to the "LifeStream Manager", a clinical dashboard, where it can be viewed by care providers, giving them access to trends, and any unusual activity.

The Genesis Touch was designed to facilitate user mobility and simple operations. Using a high contrast interface, users respond to on-screen prompts, making it useful for older adults that are not accustomed to digital interfaces.

In addition to providing questions and education tailored to the users' medical condition, the Genesis Touch facilitates simple, one-touch video visits through an integrated video application, FUZE.

The Genesis Touch offers visual and audio prompts to guide the user through the use of the device. The product is based on an off the shelf tablet that provides enhanced security and a tamper-free environment by "locking" down the operating system and only allowing applications that have been qualified and tested for the device.

INDEPENDENT SELF CARE

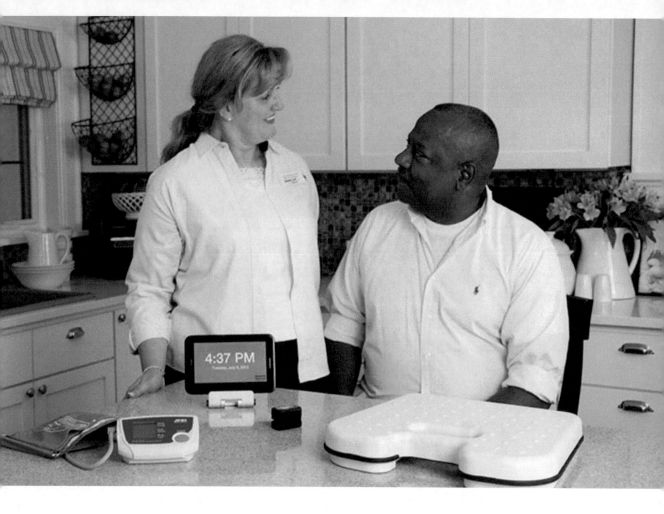

The device operates in combination with LifeStream, which provides encrypted databases that care providers can use to provide patient oversight and coaching.

The Genesis Touch uses either WiFi or the cellular network to transmit information.

MORE INFORMATION:

https://www.hommed.com/lifestream-products/genesis-touch/
http://www.businesswire.com/news/home/20140118005014/en/Face-Face-Honeywell-HomMed's-Genesis-Touch™-v2.1#.UyMNwNzi6oU
http://www.homehealthnews.org/2012/02/himss-2012-in-brief/

Content in this section provided by Stacey Force.

P

Clockreader

ELLEN YI LUEN DO AND HYUNGSIN KIM   2011

The "Clockreader" is an interactive application that allows older adults to self-administer dementia screening at their homes. It is a pen-based application modeled after the cognitive screening "clock drawing" test.

The test is based on a simple exercise, where participants are asked to use a pencil and paper to draw the minute and hour hands on a clock according to random times of the day.

Monitoring results over time allow researchers to see changes in cognitive functions. The Clockreader aims to make this data trackable in digital format and provide more precise feedback regarding spatial perception and time stamping for all information received.

The Clockreader is made using the C# programming language and Microsoft Visual Studio on an XP Tablet PC to handle the input from the pen based display. Individual strokes are recorded to be analyzed by embedded computing capabilities.

The software provides spatial, temporal and user sketch information, along with behavior data, including time required to complete the task and pressure of the pen during the exercises.

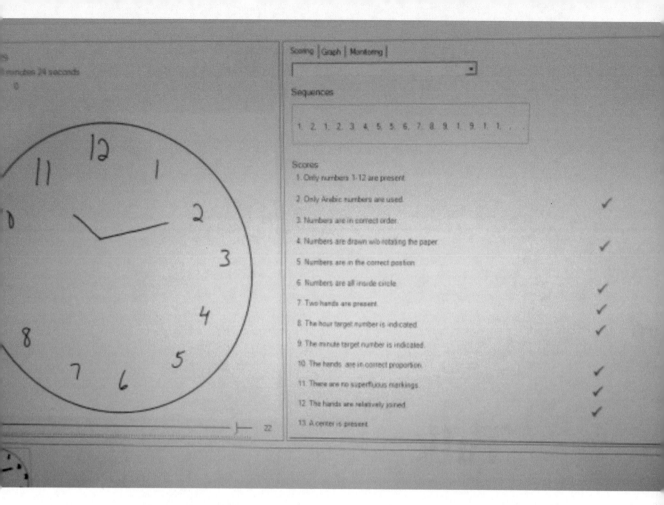

The "clock drawing" test has been administered in the same way for more than two decades. By utilizing digital technologies, the test has increased functionality without requiring an evaluator to oversee the test process.

MORE INFORMATION:
This project is supported by the National Science Foundation, Korean Institute for the Advancement of Technology, Atlanta Clinical & Translational Science Institute, Health Systems Institute, and the Alzheimer's Disease Research Center at Emory.

https://wiki.cc.gatech.edu/designcomp/images/8/89/CHIwksp–ClockReader.pdf
http://acmelab.gatech.edu/?p=10292

Credit to the photos, Georgia Institute of Technology.

Content in this section provided by Ellen Do's listed links.

DESIGNED TECHNOLOGIES FOR HEALTHY AGING   81

C

Talk

TOM HARRIES   2011

The "Talk" is a two way wearable wireless mobile device for older adults. Utilizing mobile communications networks, the device will connect older adults to care providers as well as friends and family.

The device was designed to simplify the interface for emergency situations anticipating the needs of both the target user and telecare operators.

The design features a simplified and intuitive interface that aims to increase proper usage of wireless mobile communications by those in need. By removing the functions of a cell phone, the design allows for only three calls to be made. The limited functions allow increased usability of the mobile device.

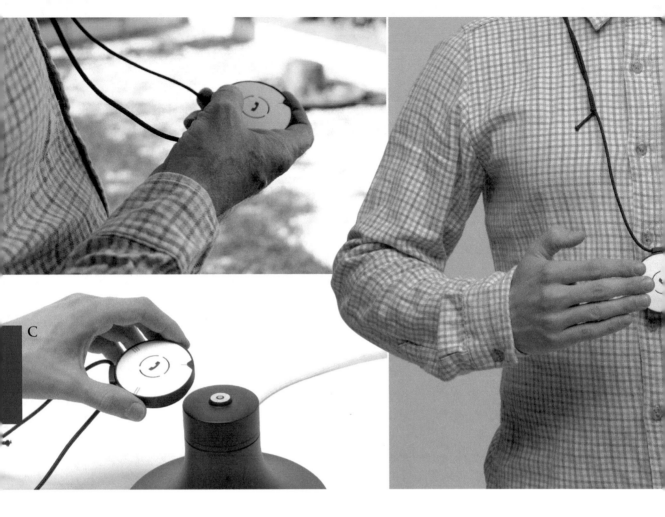

C

Talk features a front display with 3 call icons—each with programmable phone numbers. Family members, telecare operators or caregivers can be programmed and be contacted when needed. As a default, the design ensures that an outbound number is always selected.

A clip and neck strap ensure the device is front facing and easily accessible when it is needed. To make a call, the whole front face is pressed and held for 3 s. In the event of an emergency, minimal dexterity is needed to call for help. For charging, the device connects to an electrical 360° rotational base.

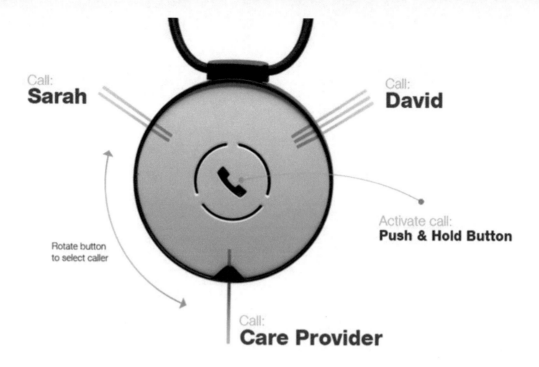

Call:
Sarah

Call:
David

Activate call:
Push & Hold Button

Rotate button
to select caller

Call:
Care Provider

Talk emphasizes the need to bring about solutions to live independently yet having fast access to help when needed. Leveraging with mobile networks, Talk is a viable product to current communications markets without jeopardizing the core function of connecting older adults in need of help.

MORE INFORMATION:
http://www.yankodesign.com/2011/06/27/medical-emergency-necklace/

Content in this section provided by Tom Harries.

P

Babyloid

Masayoshi Kanoh, Taro Shimizu and
Yukio Oida

2008

The "Babyloid" is a portable robot that helps relieve the psychological stress experienced by older adults when moving to long-term care facilities. Babyloid is designed to offer a limited number of interactions.

The interactions are based on expressing emotions such as crying, and having users respond to them with attention. It tries to comfort older adults by having a companion with expressive conditions.

The aim of the project is to ease depression among older adults and increase the feeling of purpose in life by giving them a caregiving role. Babyloid mimics the behavior of a baby including gestural expressions though blinks, smiles and even sounds. Users can also hug or rock the robot to sleep and receive interactive feedback.
The robot is a furry abstract baby size shape with a large flat surface portraying eyes and mouth.

The face is made of silicon resin 1.5 mm thick. The resin can be stretched by motors in the mouth and jaw region to create various expressions. Eyes are also rotated by motors, and LEDs are placed in the cheek regions, tears and blushing. It has awareness of the surrounding environment by using a camera, microphone and sensors.

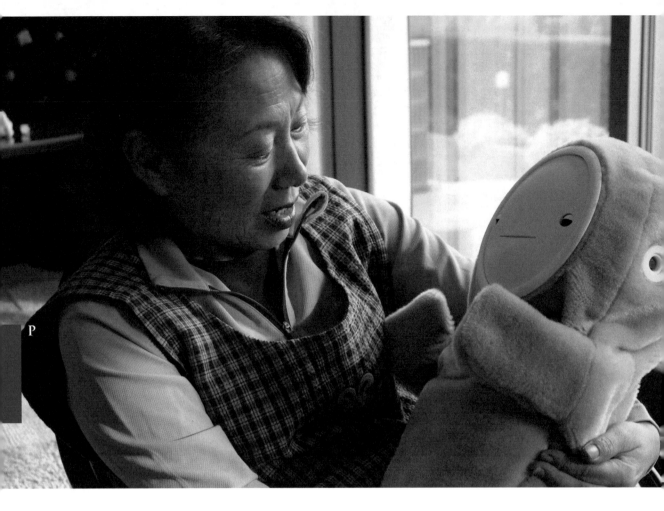

P

The robot can express feedback as voice through a speaker. The voice is from a real one year old human baby sounds. With the use of multimodal interactions, the robot allows emotionally rich expressions.

MORE INFORMATION:

Kanoh, M., and Shimizu, T. (2011). Developing a Robot Babyloid That Cannot Do Anything. Journal of the Robotics Society of Japan, 29(3), 298-305.

http://scitechdaily.com/babyloid-robot-aims-to-combat-depression-in-the-elderly/
http://www.ibtimes.com/babyloid-therapeutic-robot-can-cure-loneliness-depression-709432
http://www.st.chukyo-u.ac.jp/z104123/research.html

Content in this section provided by Masayoshi Kanoh.

Healthy

Home

It is estimated that, by 2030, older adults will represent 20% of the total population, and that by 2050, 50% of older adults who require care will not have children.[1] These statistics poses the question on how older adults can age and remain in their homes with little or no family support. Even though there are currently different housing options available for older adults, including independent and assistive living retirement communities, nursing homes, continuing care retirement communities, naturally occurring retirement communities, to mention a few, aging in place remains the ideal option for a healthy agings. But aging in place demands on the design of technologies for the home that can assist older adults aging independently and safely in their homes. Communication and sensing technologies can help older adults making them feel safe and in control of their lives. They can help establish a support network for general to health related needs. Yet, the challenging task for technology applications is to address the issues of privacy and how to integrate them seamlessly into the environment.

"Lively" presents a sensor-based system of products that can be easily inserted in current homes, providing proactive activity sharing to mobile and computing devices. Similarly, "Sonamba" non-intrusively monitors activities of daily living, including wearable devices such as panic buttons. With constant monitoring, "DigiSwitch" addresses the issue of privacy by proposing an interface application to manage access of daily monitoring to families and caregivers.

"iHealthHome" showcases a practical solution to facilitate caregiving through a comprehensive services platform. The application features services such as health tracking, video conferencing, check-in, checkout systems, to mention a few.

"Glance" features technologies for the home that have been embedded in the environment to assist older adults in tracking home resources data such as air quality or energy problems, and recommended alert system for unusual behaviors. While "WIMI-Care" and "VGo" present the emerging opportunity of implementing robots in the home to assist older adults. In particular, they address the issue of telepresence and identity building through technologies.

Overall, this section presents technologies that have been designed for the home in order to support older adult activities. Examples include technologies that monitor older adult behaviors and connects them with families and caregivers, as well as robotic interventions that can assist older adults in activities.

1. The Society of Certified Senior Advisors. (2011). State of the Senior Housing Industry, Society of Certified Senior Advisors. Denver, CO, USA.

M

Live!y

BOULD DESIGN AND LIVELY 2013

The "Live!y" is a system of activity-sharing products that support the independent lifestyle of the aging population, helping families and loved ones remain connected. The system provides proactive activity sharing using passive sensors placed around the home and mobile devices. The sensors do not require an internet connection and are placed around the home to help older adults with their everyday activity-sharing experience.

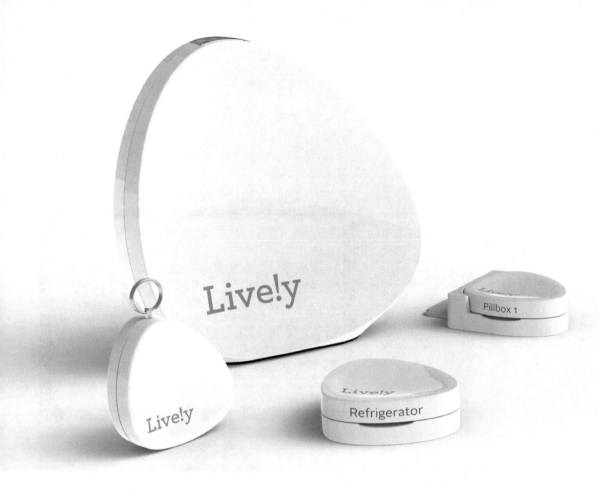

Live!y is simple to use and install, without interfering with older adults' normal activities. It offers the utility of learning about the user. The system woks by gathering information about the user. The Live!y sensors, placed on objects within the home, learn an older adult's routine for medications, food and drink, getting out, and more.

Once plugged in, the Live!y hub captures activity signals from the sensors and sends it securely to the Live!y website via a built-in cellular connection in which no home internet connection required.

Live!y

David

System status
All systems go!

Notifications
10 new updates.

Daily Routine

LivelyGram

Notifications

Circle

🦠 **Medications**
Last active: None

Edit routine   Back to at-a-glance   ⓘ

	sun	mon	tues	wed	thurs	fri	sat

Midnight

6 am   6:22
            7:0

Noon

6 pm

11 pm
            0   0   3

M

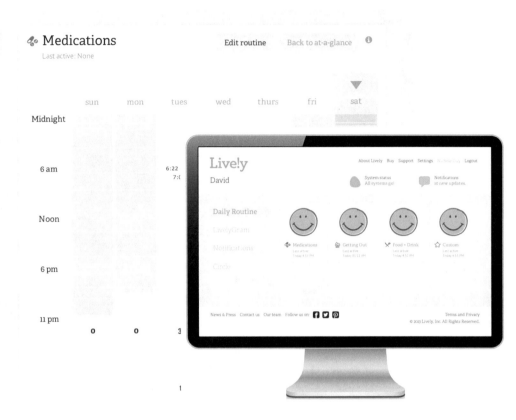

It also allows older adults who live independently to share information with their families and/or their loved ones. Older adults have the capability of sharing the collected information via web or smart phones as well as setting up notifications. The system can send notifications when activity patterns change as a reminder of daily preferences. The system also connects users.

Closing the activity-sharing loop, a printed Live!yGram mailer containing photos and messages from family and friends is automatically created twice a month for each independent older adult.

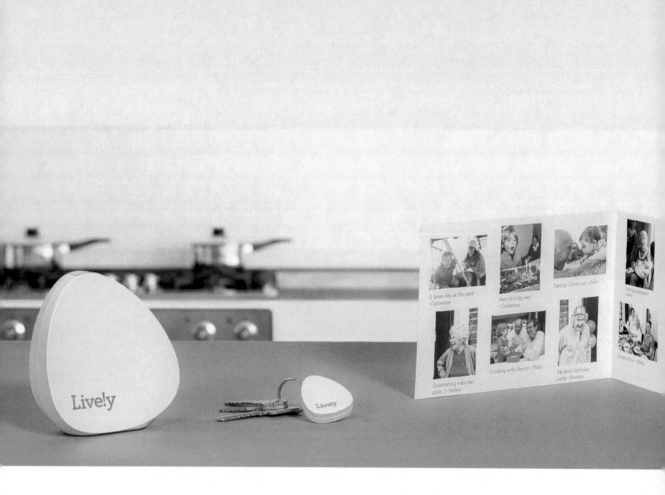

Live!y at-a-glance is focused on the well-being for everyone. The technology enables a unique activity-sharing experience between older adults and their loved ones to feel more connected to events that shape each other's daily lives. It aims at improving communication interactions that would otherwise be considered parent-sitting such as a check-in call. It enables more dignified and meaningful conversations that focus on what's important: how everyone is doing rather than what they are doing. With Live!y in place, adult children feel assured their parent or relative are doing well and as a result older adults have a sense of aging gracefully.

MORE INFORMATION:

Product designed by Bould Design and Lively. Credits to co-authors Fred Bould, Emmanuel Carrillo, Iggy Fanlo, David Glickman and Keith Dutton.

http://www.mylively.com
http://www.crunchbase.com/company/lively-formerly-hamlet
http://vimeo.com/62448218

Content in this section provided by Fred Bould.

M

Sonamba

POMDEVICES   2012

The "Sonamba" is a wellbeing status monitor and medical alert system for older adults living independently. Sonamba periodically sends senior's wellbeing status alerts to caregivers and includes 24/7 personal emergency response call center, medication adherence and social communications functionality. Sonamba is designed with a "coolness" factor and lifestyle-enhanced image for the older adult group to avoid having typical medical-looking devices or emergency response systems in the home.

Sonamba is packaged as a user-friendly, aesthetically pleasing digital photo frame with touchscreen interface and built-in cellular connectivity. It is designed to be a part of everyday living — empowering seniors as well as their caregivers to live life on their own terms.

Sonamba includes comprehensive senior-oriented features: activity of daily living monitoring, personal emergency response, medication and calendar reminders, and social communications and games.

M

Sonamba non-intrusively monitors activities of daily living. The system works by placing the main device in the area of highest daily activity, such as the living room or kitchen. Additional activity sensors can be placed in other areas of the home, including bedroom and bathroom, offering monitoring of all relevant areas of the home.

Sonamba monitors motion and sound activity and compares current activity with historical activity patterns. Based on this comparison, Sonamba sends out periodic "All is Well" or "Attention Needed" alerts to caregivers and support circle members cellphones.

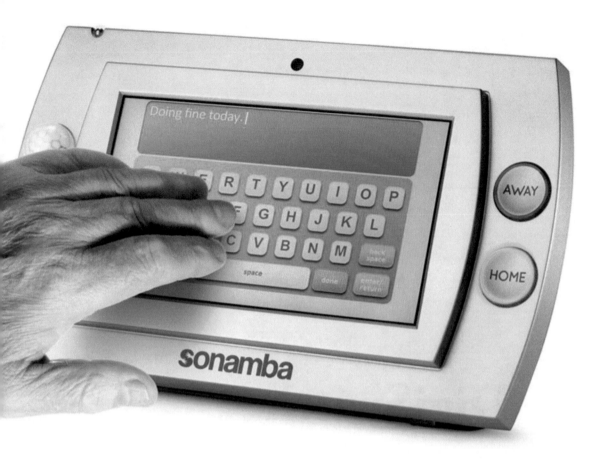

Sonamba features a panic button on the unit and one that can be worn as a pendant and wristband, or optionally, attached to the wall where a fall may occur, such as a bathroom or kitchen. In the event of an emergency, older adults simply press the panic button in order to call up to three designated caregivers.

Sonamba has its own phone number, so older adults can send and receive text messages and emails from their homes. In addition, the "Instant Notes" feature makes it easy to send status updates without having to type a new message each time.

MORE INFOMRATION:

Sonamba Monitoring System was a nominee in the Best Senior Living Awards 2013 in the Most Innovative Products.

http://sonamba.com
http://seniorcarecorner.com/technology-review-sonamba-wellbeing-monitor
http://www.zdnet.com/blog/healthcare/sonamba-takes-mass-market-approach-to-elder-care/4105

Content in this section provided by http://sonamba.com/about/press-kits/, Robert Burke.

Data Managment, Graphic Interface Design, Touch Screen Interfaces, Video Communications

P

DigiSwitch

KELLY CAINE AND CELINE CHHOA   2010

The "DigiSwitch" is a central control panel that integrates and manages home monitoring technologies within an older adult's home. It is a touchscreen computer that takes the form of a digital picture frame for use in the home.

Users can make use of DigiSwitch to turn monitoring devices on or off with the flick of a digital switch. It allows users to retain complete control over the information gathered by the various home monitoring devices.

Home health monitoring represents an appealing alternative for older adults considering out-of-home long-term care. In addition to older adults' preference for familiar surroundings, living independently is considerably more cost-effective for individuals, their families, and society, than institutionalized care. However, introducing health monitoring into the home environment poses significant privacy concerns. DigiSwitch addresses these concerns by putting users in control of their privacy. The system monitors and directs the collection and transmission of health information, allowing users to both benefit from home monitoring technology and maintain privacy.

DigiSwitch is a touchscreen PC, which also acts as a digital picture frame. The interface is desgined to be used without needing any particular technological expertise. It utilizes a Start-Stop-Pause naming scheme for the buttons, which reflects the familiar model from other technologies likely to be familiar to older adults (e.g., VCR). Using a stoplight metaphor, the color of each button changes according to the mode; green corresponds to Start, red to Stop, and yellow to Pause.

A key feature of DigiSwitch is the Pause function which allows users to pause the home monitoring device(s) so that no additional real time information is collected and transmitted to the caregiver without the caregiver knowing the device has been paused. DigiSwitch also includes a "Friend's View" screen which shows users what their caregiver is seeing. This allows users to be more in control over what data a caregiver would be able to access.

HEALTHY HOME

As the aging population continues to grow, supporting independent living in later life is a critical concern for older adults.

DigiSwitch has the potential not only for individuals wishing to age in place, but for other community living arrangements.

MORE INFORMATION:

This project is a joint collaboration of Dr. Kelly Caine, Dr. Kay Connelly, Dr. L. Jean Camp, Dr. Lesa Huber, Dr. Kalpana Shankar, Celine Chhoa, Zachary Schall-Zimmerman and Dr. William R. Hazlewood.

ETHOS Living Lab: http://ethos.soic.indiana.edu

This material is based upon work supported by the National Science Foundation under award number 0705676. Any opinions, findings, and conclusions or recommendations expressed in this presentation are those of the author(s) and do not necessarily reflect the views of the National Science Foundation.

Caine, K. E., Zimmerman, C. Y., Schall-Zimmerman, Z., Hazlewood, W. R., Camp, L. J., Connelly, K. H., Huber, L. L., & Shankar, K. (2011). DigiSwitch: A Device to Allow Older Adults to Monitor and Direct the Collection and Transmission of Health Information Collected at Home. Journal of Medical Systems, 35 (5), 1181-1195.

Content in this section provided by Kelly Caine and Celine Chhoa.

M

iHealthHome®

DEW-ANNE LANGCAON   2009

The "iHealthHome" is a comprehensive in-home monitoring, collaboration and documentation system designed by professional caregivers for professional caregivers. The system allows older adults to age in place and remain independent and self-sufficient. iHealthHome is a dynamic platform that coordinates interaction and feedback between older adults at home and their care team including professionals, family and service providers. The goal of iHealthHome is to enable older adults to "Live

Well at Home" by promoting: self-management of health conditions; remote monitoring for activities in the home; electronic documentation and collaboration among the caregiving team; efficiency and cost reduction of in-home caregiving; and transparency and participation of family members in planning for care and services for older adults remaining at home. The device interface is an easy to use touchscreen monitor that is installed in the home of an older adult. Wireless devices such as motion sensors, blood pressure monitors, glucometers and scales can be added to monitor older adults and their environment. The iHealthHome platform operates alongside professional applications to help caregivers communicate and collaborate.

iHealthNavigator is an electronic medical record for registered nurses and care coordinators for tracking and reporting information. This data enables preparation of assessments and care plans that are displayed on the touchscreen in the home.

Flowsheets is a simple to use electronic documentation system by exception for in-home

caregivers to legibly organize notes, interventions and outcomes in a digital format.

Clock In/Clock Out is a time and attendance system that helps with timekeeping for payroll and billing.

Services at Home is a virtual concierge e-commerce application that is a cashless ordering system

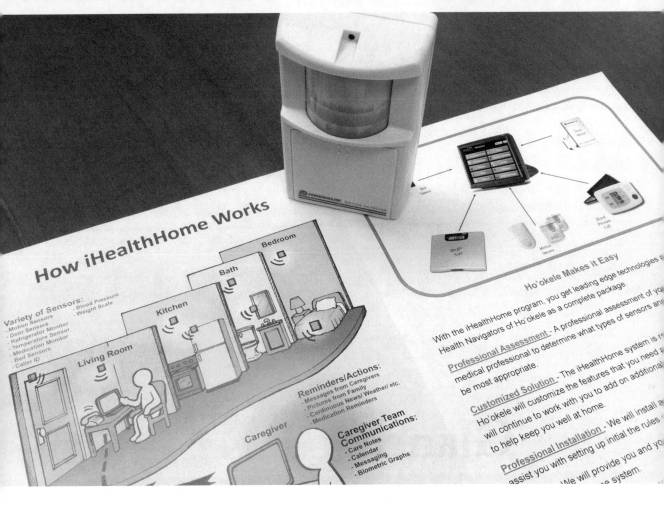

for groceries, taxis, prescription refills, etc.

The Family Portal is a communication tool that empowers families to collaborate with caregivers.

iHealthHome enables all members of the team to stay connected to older adult's activities at home and to each other with the goal of keeping them safe, healthy and independent at home.

MORE INFORMATION:
Co-authors include Bonnie Castonguay and Phill Moran.

www.ihealthhome.net
http://www.healthcareitnews.com/news/hawaii-hit-pilot-sees-early-success
http://www.bizjournals.com/pacific/print-edition/2010/10/22/health-care-coordinators-turn-to-high.html

Content in this section provided by Dew-Anne Langcaon.

Embedded Computing, Graphic Interface Design, Sensing technologies

Glance

LARRY FREIL, MUDIT GUPTA, JAMES HALLAM, NITYA NORONHA, LAUREN SCHMIDT AND PHILIP SMITH   2013

The "Glance" is a home health wellness dashboard and interactive smart mirror designed to manage resources in the home. Glance keeps track of five separate resources in the home including plants, electricity, water, gas and air, and notifies or alerts users of significant changes. When Glance has no notifications, the interactive system fades into the background and functions as a mirror.

Older adults spend a good amount of time in their home performing everyday activities. The home

plays a central role in older adult's health, lifestyle, and well-being. Yet, there are necessary home resources that need particular attention to avoid endangering the life of older adult's aging in place. Facilitating monitoring activities can become not only an alleviation for older adults caring for the overwhelming needs of their home but also a factor to age in place.

Glance is an interactive system comprised of a mirror and sensing technologies designed to monitor and alert users of various resources at the home. The system can monitor plants, electricity, water, gas and air.

The mirror is hung in a central location, and connected to sensors around the home. The Glance kit also includes wireless soil moisture sensors and air quality sensors. Glance monitors the sensors for changes in the resources Glance recognizes when users pass by, and flashes notifications through icons about the home resources.

Each resource has an icon with a unique color and position in the mirror. Each icon can animate— sliding in up and down arrows to show changes in consumption, and smiley faces to indicate changes in quality. The level of notification changes in intensity based on recent changes in the individual nutrient data streams.

HEALTHY HOME

GLANCE

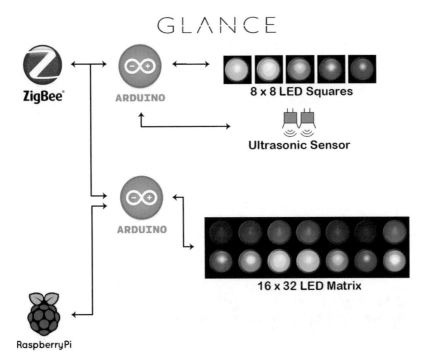

Glance communicates using Zigbee wireless technology, getting inputs from the sensors around the home. Glance uses an ultrasonic and passive infrared sensor mounted on the frame to detect when a person is standing in front of it, and displays notifications using a set of LED matrices.

MORE INFORMATION:

Glance is a project as part of the Aware Home Research Initiative (AHRI) at Georgia Institute of Technology is an interdisciplinary research endeavor aimed at addressing the fundamental technical, design, and social challenges for people in a home setting. Central to this research is the Aware Home, a 3-story, 5040 square foot facility designed to facilitate research, while providing an authentic home environment: http://www.awarehome.gatech.edu/#sthash. UV3MpeRW.dpuf.

Glance won second place on the Connected Home Convergence Innovation Competition at the Georgia Institute of Technology: http://cic.gatech.edu/spring-2013/winners

Content in this section provided by James Hallam.

P

WIMI-Care

Silvana Cieslik and Peter Klein   2011

Within the research project "WIMI-Care", a robotic system was developed to assist nursing staff by taking over routine tasks and allowing them to spend better quality time with older adults.

The system is comprised of two robots: Care-O-bot and CASERO. The robots offer different services and interactions for older adults and their caregivers.

The services and interactions were designed to meet the needs and everyday requirements of nursing home staff and residents. Some of the routine activities supported by Care-O-bot and CASERO include reminders, emergency alerts and assistance, transportation of items, games and memory training.

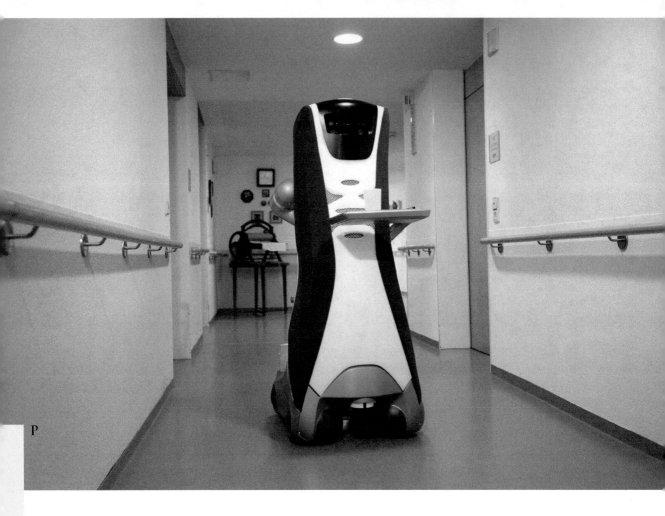

Care-O-bot and CASERO are connected with tablet devices to support communication with caregivers. In case the robots recognize unusual behavior, they send a message to the caregiver or nursing staff's tablet. Likewise, the caregiver can call over the robots for help in case of emergency.

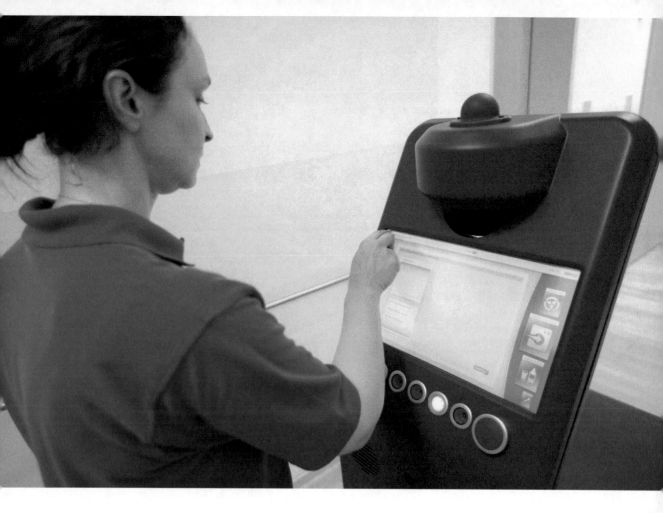

In addition, the robots can be controlled directly, by using a PC or a tablet device. The interfaces were designed with a consistent operational concept and screen design, which allows staff to operate the robots intuitively.

MORE INFORMATION:

Project partners include Fraunhofer Institute for Manufacturing Engineering and Automation (IPA), MLR System GmbH, University of Duisburg-Essen and User Interface Design GmbH (UID).

https://www.uni-due.de/wimi-care/index_en.php

Content in this section provided by Kathrin Schwarze.

M

VGo

VGo Communications Inc.

2008/11

The "VGo" is a telepresence device that allows people to locate and communicate with older adults via a mobile avatar. VGo replicates a person in a distant location.

The device features two-way audio and video plus remote controlled movement in order to emulate a face-to-face conversation.

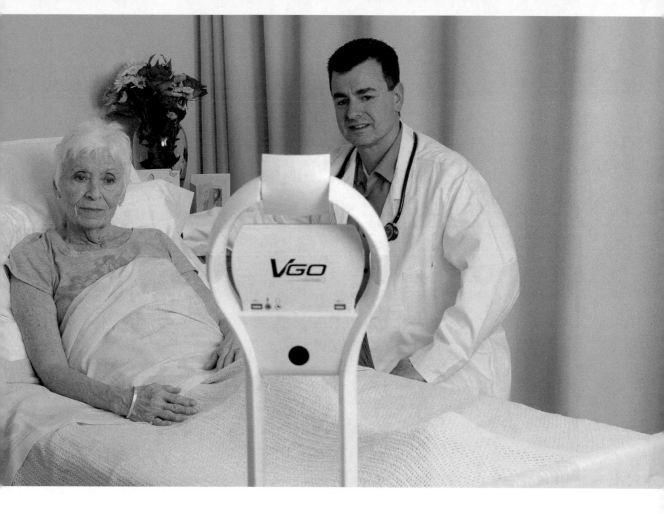

The solution comprises two primary elements: a remote controlled mobile device the represents a person in a distant location; and the software application that is downloaded to a computer.

The application is used by a remote user to initiate connectivity, see and hear the remote locations. In addition, it also controls the telepresene wheeled device. Users can navigate remote locations and and interact with other people as they would if they were there in person.

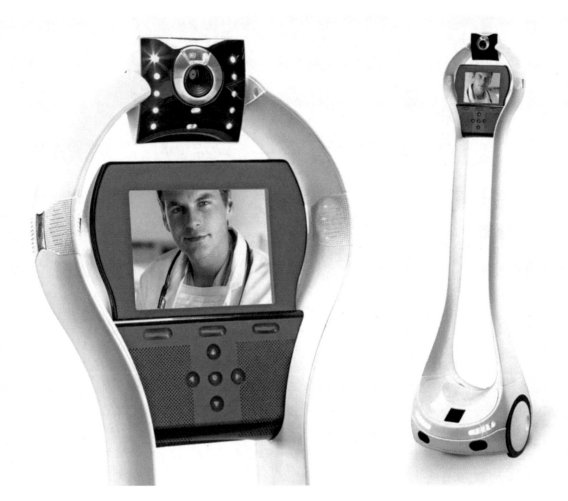

VGo is currently in use in assisted living communities, long-term rehabilitation facilities, and nursing homes to empower disabled and older adults to live more independently. VGo can be used by family members to visit from a distant location.

VGo is also used in the home to provide immediate access and medical advice by healthcare staff and family. This lowers the barrier between caregivers and those who need minor medical attention and could lower the global cost of healthcare.

VGo connects to its remote user with Wi-Fi or Verizon 4G LTE. The solution does not require any knowledge by the person interacting directly with the VGo since it is remote controlled by a distant user.

MORE INFORMATION:
http://www.vgocom.com
http://www.huffingtonpost.com/tag/vgo-robot
http://www.psychologytoday.com/blog/media-spotlight/201306/can-robots-help-care-the-elderly

Content in this section provided by Angela Paris and Ned Semonite, VGo Communications Inc.

Active

Lifestyle

As older adults age, remaining active is a necessary measure for healthy aging. The Centers for Disease Control state that promoting physical activity among older adults is a national public health priority.[1] They suggest that the regular physical activity substantially delays the onset of functional limitations and loss of independence, reduce the amount of cognitive decline, and often reduces symptoms of depression. Being active is not only about benefiting from physical activity but also also considering how older adults engage in activities that help them keep healthy. Successful aging is largely determined on an individual's active lifestyle including the body, the mind and connections to communities. Remaining actively connected with others can help older adults adopt choices for a healthy lifestyle. The next pages portray exciting products that represent opportunities for active lifestyles of older adults.

"The Aid" is designed to not only help older adults during rehabilitation, but mainly to give them confidence and that they are not alone in the process. Likewise, "Social Yoga Mats" are interactive products to facilitate exercise among a group, especially when you are located at a distance. The main focus is that activity is associated with group support. "el: Dudy" represents promising exercise products working with sensing technologies that accurately record physical exercise data for the user. This project is an example of emergent products designed for aging based on awareness as a motivation factor. Similarly, "Manea" is designed for cognitive exercising aimed at preventing cognitive disabilities.

New interactive games also form the plethora of products targeted to keep older adults active. "Memory Monopoly" offers a flexible reminiscing rehabilitation game with interactive activities, especially aimed at older adults with dementia. Both "TEPOS"

and "YoooM" portrays products designed to enhance social connectedness through gaming activities through offering natural modes of communication interaction.

Social activities are equally important for active lifestyles. "Table Talk" and "Photostroller" represent products designed to encourage communication interactions among older adults in community living environments. Lastly, "Prayer Companion" presents a simple technological interventions that can enhance the sense of purpose among older adults.

From smart canes to interactive tables, these designed technologies exemplify the necessity to stay active at various levels, physically, cognitively and emotionally, but especially focusing of cooperative support.

1. Centers for Disease Control and Prevention (2014). Promoting Active Lifestyles Among Older Adults. From http://www.cdc.gov/nccdphp/dnpa/physical/pdf/lifestyles.pdf.

The Aid

C

Egle Ugintaite

2010/11

"The Aid" is designed for older adults and patients who have suffered trauma. Older adults and recovering patients often lack confidence to leave their home environment after the trauma, leading to isolation, depression or even danger.

The Aid takes the form of a support cane with embedded sensors and tracking tools that help users in the recovery process. A GPS communicates with a remote service to guide the user and prevent them from being lost.

The Aid combines new computer technology solutions with established cane-like function and combines physical and virtual interaction to provide full and versatile support. Embedded sensors monitor users' health data from blood vessels in the wrist including temperature, pulse, and blood pressure.

In a case of emergency, the product enables the user to call for help by simply pressing an SOS button. When pressed, the user is connected to the help center. The user's current health data and location are sent to the help center for assistance in an emergency.

C

Designed with universal principles in mind, the advanced technologies embedded in a simple form allow for intuitive use. The Aid is designed to be a smart, reliable, attractive and considerate assistant which gives the user options in navigating environments, despite physical weaknesses or deteriorating health. After communicating their destination, the user is led safely via step-by-step instructions that are sent directly to the device. These directions are then communicated as audio to the user through wireless headphones included with the product.

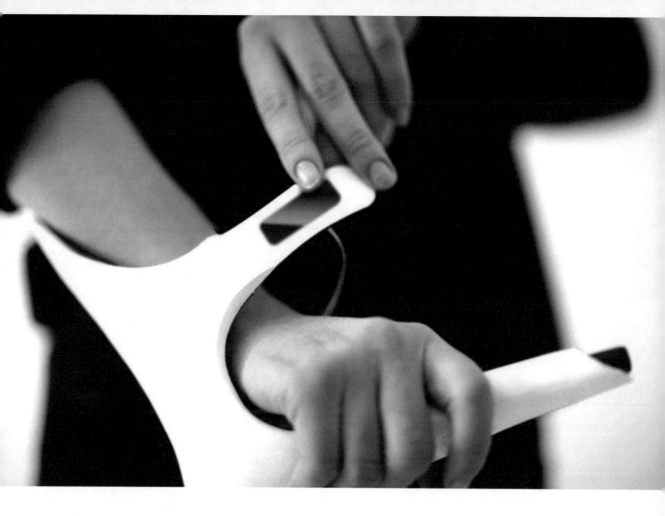

The Aid aims at helping older adults to become active members of society as well as enabling and encouraging safe mobility by providing physical and emotional support.

MORE INFORMATION: READING

The Aid received the Fujitsu's 2011 design award: A Life with Future Computing.

http://www.springwise.com/health_wellbeing/theaid/
http://health.infoniac.com/smart-cane-watches-over-your-health.html
http://www.trendhunter.com/trends/the-aid-cane
http://www.tomsguide.com/us/Aid-Cane-Navigation-Health-System,news-11277.html
http://www.dexigner.com/news/23104

Content in this section provided by Egle Ugintaite.

Embeeded Computing, Graphic Interface Design, Internet Connected, Sensing Technologies, Social Media, Wireless Technologies

Social

Yoga Mats

P

KARL MAYBACH, ARUN NAGARGOJE AND TOMAS SOKOLER MEDPLAST

2011

The "Social Yoga Mats" is an interactive system designed for older adults. The system provides automated yoga instructions and facilitates social sharing of exercise data among members of a peer network formed during group yoga classes. The design provides each user with a yoga mat and a portable touch-screen display, which together enable users to notice each other's exercise activities while also being noticed for their own efforts.

The design builds on group motivation and the relationship

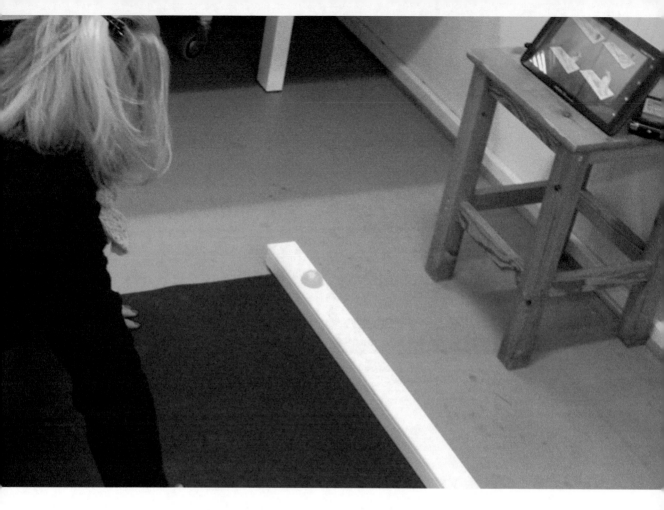

between exercising and socializing. It aims to extend these mutual benefits from weekly group yoga classes to solitary exercise at home. As a long-term solution, the design seeks to motivate older adults to maintain regular exercise while strengthening existing relationships.

The design consists of a set of exercise mats with embedded pressure sensors, activity indicators and wireless internet connectivity, networked touch-screen tablet devices, and a central server and database. Each mat detects the presence of a user performing yoga exercise and registers this activity with the central server.

Video lessons and real-time peer activity indicated by cartoon avatars.

Motivate peers by sending awards and applause for their activity.

P

The Social Yoga Mats aim to enhance the physical fitness and social wellbeing of seniors, both of which are challenges and essential elements for healthy aging.

The design provides concrete support for exercising through video instruction while also strengthening the social network built around a mutual interest in yoga.

Each tablet device provides video yoga lessons drawn from weekly classes and also displays members of the peer network using yoga-themed cartoon avatars. The tablet shows the real-time activity status and exercise history of each peer and delivers achievement awards. Simple messaging features allow for greater interactions such as sending health tips, applause and awards to the peers.

Using the tablet interface, peers can access detailed information and begin to notice the exercise rhythms and routines of others. Having mutual awareness encourages participation and increases social connectedness among the group members. Having a support system of physically healthy peers provides opportunities to socialize and offers motivation to continue exercising.

The system builds on existing as well as new social networks. It provides control over the degree of one's exposure, alleviating privacy concerns.

The form of social sharing through fitness facilitates and strengthens social ties among yoga group members over time. Through this product, older adults are encouraged to work out and interact with others with mutual interests, addressing in an unobtrusive manner the issue of loneliness.

MORE INFORMATION:
Design exploration undertaken for project Lev Vel: http://www.lvvl.dk.
Project developed at the IT University of Copenhagen.

Nagargoje, A., Maybach, K., & Sokoler, T. (2012). Social Yoga Mats: Designing for Exercising/Socializing Synergy. Proceedings from TEI'12: The Sixth International Conference on Tangible, Embedded and Embodied Interaction. Kingston, Ontario, Canada.

Maybach, K., Nagargoje, A., & Sokoler, T. (2011). Social Yoga Mats: Reinforcing Synergy Between Physical and Social Activity. Proceedings from CHI '11: Conference on Human Factors in Computing Systems. Vancouver, BC, Canada.

Content in this section provided by Arun Nagargoje.

el:Dudy

CHAN PO YEE 2010

The "el:Dudy" is a handheld home exercise system designed for older adults. The system is comprised of three differentiated exercise tools. Each tool is equipped with motion sensors and accelerometers to track use.

All the information is synchronized and stored on personal computers or smart phones via Bluetooth. The system aims at facilitating exercise at the home while also motivating older adults to exercise through awareness.

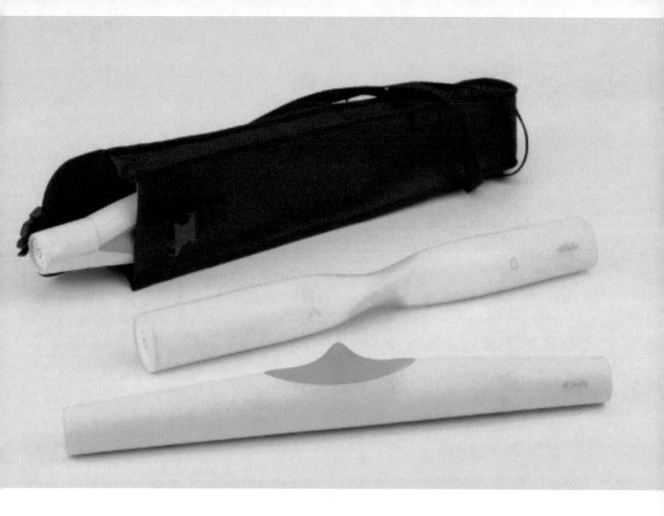

Exercising is a challenge among older adults. Lack of exercise can have an impact on health with diversified complications, among them losing muscle mass and strength. Even though there are a number of programs and equipment for exercise, none addresses the issues of exercising in small spaces and the comfort of one's home.

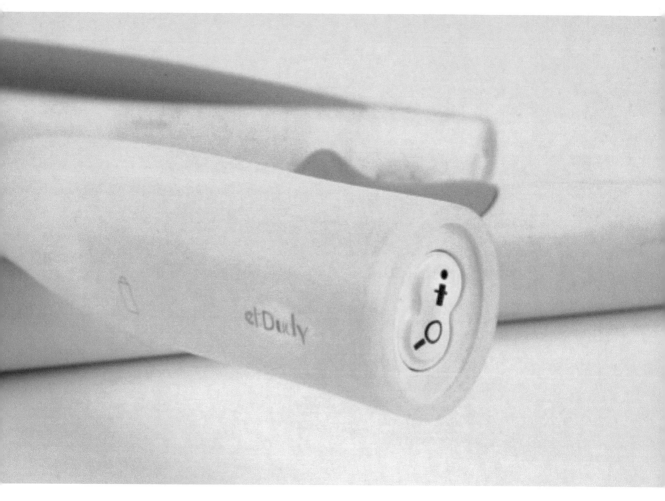

C

el:Dudy is an all in one handheld exercise system. The design is based on bare gymnastic exercises that allow users to do diversified exercises to train up the human body comprehensively, leading to a richer exercise than that of bare gymnastic exercise.

The device allows different exercises, including doing wrist-twists, stretching or giving massages. Each exercise is linked to a physical design. The sensors give accurate readings by utilizing 3D motion sensors that detect precise counts, motion acceleration, exercise duration and strength.

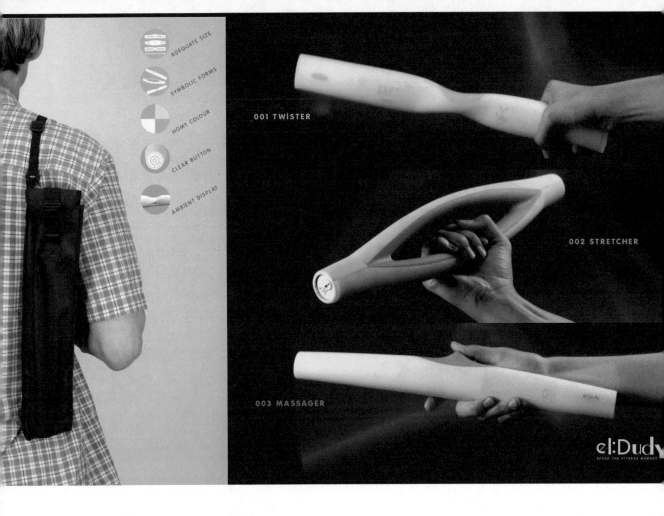

These tools were designed to facilitate exercise and avoid hard-to-do exercises. The goals were to design tools that promote active aging toward improving quality of life and life expectancy for all people as they age.

MORE INFORMATION:

Project developed undet the guidance of Mr. Benny Leong Ding, Hong Kong Polytechnic University.

http://www.yankodesign.com/2010/11/03/eldudy/
https://www.youtube.com/watch?v=EildSBwKuBs
http://www.ubergizmo.com/2010/11/eldudy-handheld-exercise-system/
http://content.yudu.com/Library/A1xc2j/ResearchProposalthea/resources/2.htm

Content in this section provided by the listed links, Chan Po Yee.

Manéa

C

LUIGI TRABUCCO 2007

The "Manéa" is a multimedia-based training device that boosts mental ability and helps prevent cognitive disabilities. It has a series of games and exercises to train different areas of the brain. Its training exercises require the use of both hands and also develop both sides of the brain. The main screen is a curved OLED display with audio feedback that gives directions and hints. The side grips are rubberized for soft handling and long term comfort during activities. The input for interaction is located in a spherical shape,

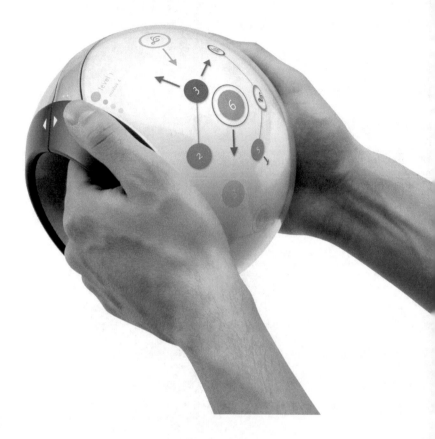

which is a soft surface that is pressed to progress through the activities. As the user progresses, the device will recognize and learn to strengthen weak areas of learning or motor control.

Navigation is managed through an embedded accelerometer as well as by rotating the side grips. The device is designed to receive information from bluetooth and wireless access points.

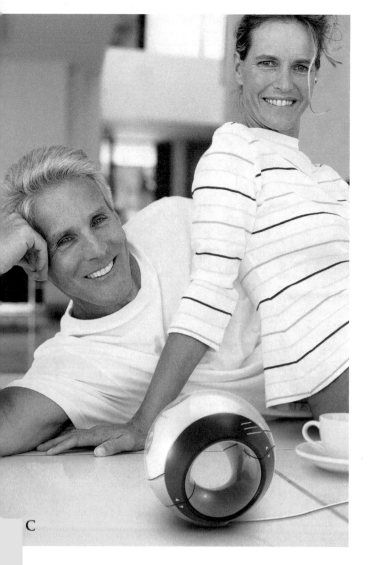

C

switch between
the answers

take up and release
the single parts

press the wanted finger
in the foam with
the demanded pressure

the more sure you are,
the more you
have to confirm

move the single parts
in the area

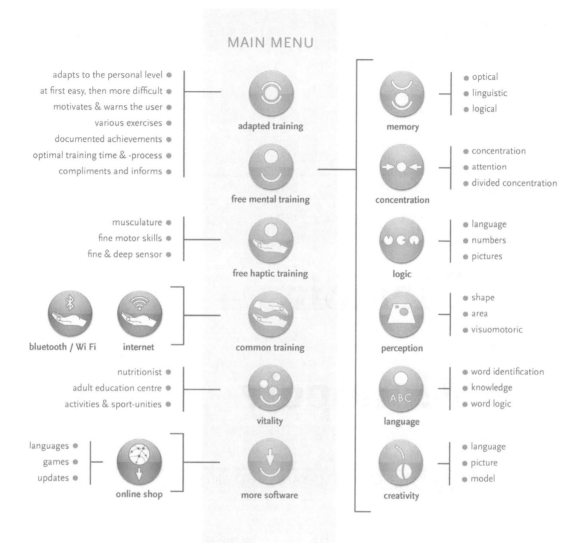

MAIN MENU

adapts to the personal level
at first easy, then more difficult
motivates & warns the user
various exercises
documented achievements
optimal training time & -process
compliments and informs

adapted training

free mental training

memory
- optical
- linguistic
- logical

concentration
- concentration
- attention
- divided concentration

musculature
fine motor skills
fine & deep sensor

free haptic training

logic
- language
- numbers
- pictures

bluetooth / Wi Fi   internet

common training

perception
- shape
- area
- visuomotoric

nutritionist
adult education centre
activities & sport-unities

vitality

language
- word identification
- knowledge
- word logic

languages
games
updates

online shop

more software

creativity
- language
- picture
- model

Designed for older adults, the
device promotes healthy cognitive
and motor function for users of
any age.

MORE INFORMATION:
*This concept has won awards including: 2010 Red Dot Design Concept Award, Second Round 2007 Braun Prize,
Special Recognition 2007 Lucky Strike Junior Designer Award and First Place 2007 Pininfarina Design Award.*

http://www.luigitrabucco.com

Content in this section provided by Luigi Trabucco.

Graphic Interface Design, Mobile Applications, Tangible Interfaces

Memoir

Monopoly

P

HSIEN-HUI TANG, SZU-YANG CHO AND
YA-FANG CHENG
2014

The "Memoir Monopoly" is a rehabilitation game application designed to engage older adults living with dementia in rehabilitation activities. Memoir Monopoly offers a flexible reminiscing rehabilitation game with interactive activities, including structured reminiscing, cognitive training, reality orientation, sensory stimulation, and social events. The game application was conceived working with occupational therapists. Rehabilitation tools for older adults living with dementia

are needed for coping with the condition. Most paper-based rehabilitation tools have the drawbacks of inflexible designs and of limited ability to stimulate the user. Based on user experience research conducted by the design team, it revealed the need for a more interactive reminiscing game.

P

Memoir Monopoly uses four tablet devices (iPads) by synching and sharing screens. Tangible "tokens" are used for physical indications and interactions. The tokens portray the user faces on the surface. Each iPad is used to collect user's personal photos and preferences. When placed together, the individual contents are pulled together. During group reminiscence rehabilitation sessions, occupational therapists can responsively create unique games (i.e., maps) adapting the level of difficulty to suit the user's personal experiences, increase their interest, and encourage their recall of the past. Various cognitive stimuli, including watching movie clips, listening to favorite songs, and touching interactive games help participants to experience a sense of accomplishment and satisfaction in the rehabilitation process.

Utilizing a game that matches their life experience and various interactions can encourage older adults to participate in reminiscence activities as well as improving their quality of life by engaging in meaningful group activities.

MORE INFORMATION:

Project developed at the National University of Taiwan of Science and Technology, National Taiwan University, Taipei Nan-gang Seniors Service Center, Taipei Zhongshan Seniors Service Center, Taipei Aiai Retirement & Assisted Living Facility and Jianshun center. Credits to: Szu-Yang Cho, Ya-Fang Cheng, Hou-Ren Chen, Wei- Chen Chu, Mike-Y. Chen, Hung-Hsuin Ko, and Chien-Hsiung Chen, Hsien-Hui Tang.

Cho, S-Y., Cheng, Y-F., Chen, H-R., Tang, H-H., Chu, W-C., Chen, M-Y., & Chen, C-H. (2014). Memoir Monopoly: A Rehabilitation Game for Elderly Living with Dementia. Proceedings from ISG'14: The 9th World Conference of Gerontechnology. Taipei, Taiwan.

http://ditldesign.sqsp.com/design-projects/#/2014-memoir-monopoly/
http://vimeo.com/86666282

Content in this section provided by Hsien-Hui Tang.

TEPOS

C

SIQI LIU 2012

The "TEPOS" is a tangible entertainment projection system for older adults. The system makes use of tangible user interfaces bridging the digital world and the real world. Older adults manipulate the system's digital information through physical objects. The goal was to design a system that enhances social connectedness by offering an enjoyable, engaging, and mentally/physically sustainable natural method for social interactions. Although other gaming systems such as the Wii are popular, it's

not originally designed for older adults in terms of game contents and game control. As such, these systems compromise an effective active participation of older adults in game activities. There is a need to involve older adults in activities that can be a source of fun and entertainment, and at the same time improve their physical and psychological well-being.

TEPOS system was designed to address the needs to offer older adults an enjoyable way of spending time, have an appropriate interface, maintain mental and physical health and enhance seniors' social connectedness.

TEPOS system can provide older adults play and entertainment that can engage them in varied activities.

The system design looks beyond the traditional perspective of usability requirements imposed by age-related functional limitations.

TEPOS system makes use of projection technologies in order to display digital games. It also offers engaging content combined with tangible interfaces that can easily and pleasurably be used. Users can push a tangible "block" up and down to adjust the volume or rotate the block to draw shapes.

While testing the design, older adults confirmed the design considerations. They found themselves engaged in playing all games provided. The tangible manipulations were appropriate and intuitive to use.

MORE INFORMATION:
http://www.tuvie.com/tepos-tangible-entertainment-projection-system-for-elderly-people/
http://vimeo.com/27776657

Content in this section provided by the listed links.

YoooM

ROBBERT SMIT

2012/13

The "YoooM" is a video mediated communication tool that displays full body gestures between users in addition to face and audio capture. While modern communication devices offer multiple applications for communicating at a distance, they often reduce the communication space. Eye contact, body language and visual cues are powerful details that bring closeness and enhance human communication.

P

Utilizing dual cameras, one pointed at faces and the other at upper torsos, YoooM enables the receiving user to interpret non-verbal cues, adding to the benefits of face-to-face interactions. The receiving user visualizes the communication through two screens of 21 inches, arranged at an angle of 135°.

This setup creates a larger, near the life-sized view of the dialogue partner. The touch panel on the lower screen allows for direct interaction with the device without using a mouse or keyboard. A high-fidelity microphone and a set of speakers, positioned at the left and rights side of the device enable a greater sound quality.

YoooM affords three interaction options.

The Meet format allows one-to-one communication, showing the communication partner to be nearly life sized.

The Club format supports communication between groups of people, up to five, and offers interactive social activities such

as playing games or surfing the Internet.

The Classroom format enables voluntary activities, offering the opportunity of sharing knowledge or interests together with others.

This system was especially designed for older adults. A population who might be at risk for social isolation, YoooM allows

P

ACTIVE LIFESTYLE

for an accessible communication and interaction over distance. It provides a high-quality face-to-face like communication experience supporting increased social presence.

YoooM was developed within a user-centered design approach. Older adults were involved in the development process through hands-on sessions. Older adult's feedback was instrumental in early development stages in order to meet the target population needs.

MORE INFORMATION:

The YoooM system was developed within the ConnectedVitality project, under the AAL Joint Program, call2: www.connectedvitality.eu. Co-authors include Katja Neureiter, Christiane Moser, Omar Jimenez and Andre Hermsen.

www.yooom.com

Vajda, L., Tóth, A., Hanák, P., Achilleos, A. P., Mettouris, C., Papadopoulos, G. A., Neureiter, K., Rappold, C., Moser, C., & Tscheligi, M. (2013). A Health-Based Use Case of the Connected Vitality Project: the Yooom in the Sterile Room. Proceedings from eGeH'13: The 10th International Meeting on eGovernment & eHealth. Desio/Monza & Brianza, Italy.

Content in this section provided by Robbert Smit.

Design Research, Embedded Computing, Sensing Technologies, Universal Design

Table Talk

P

John Davis and Michael Spear   2013

The "Table Talk" is an interactive table designed to encourage communication interactions among older adults in community living environments. Even though older adults can be physically located next to each other in community environments, little or no active interaction occurs. Table Talk addresses the issue of social disconnectedness by designing a technology intervention to build meaningful and active relationships between its users.

Table Talk is a table with projection capabilities. It operates as a conversation prompter by showcasing images, words, weather information and news to motivate older adults to establish conversations. It also uses sensing technologies to encourage such interactions.

Table Talk uses motion detection and capacitive sensing to detect its surroundings. These sensors are used to control the LED strips and projectors embedded in the base of the table. These technologies work together to create an effortless experience for the user.

Pulsing LEDs are activated by passerbys and draw interest to the table.

When users sit at the table the projector is activated and content is displayed.

When users leave the table it returns to a passive state.

P

When users are walking around in the area, the sensors trigger the LEDs to emit a warm pulsing glow. When the user sits at the table the capacitive sensors are activated and the LEDs will hold a continuous glow while the projector is activated and the slide show begins.

When users leave the table the slideshow will deactivate, and the table will return to a passive state.

Table Talk was designed by observing and interacting with older adults at retirement communities. It is designed to be accessible for all users, especially considering older adults' needs.

ACTIVE LIFESTYLE

The table dimensions were based on ADA regulations for wheelchair clearances in mind. The height easily allows enough room for a wheelchair to pass underneath. The form of the table was also defined to allow for easy access to the table from all sides with minimal chance of tripping on or hitting the table.

Future development of the table looks to implement RFID technology to tailor content to specific users allowing them to share with distant loved ones.

MORE INFORMATION:

This project was developed in the School of Industrial Design, Georgia Institute of Technology.

http://seniorplanet.org/is-this-the-future-of-design-for-aging/

Content in this section provided by John Davis and Michael Spear.

Photostroller

P

INTERACTION RESEARCH STUDIO, GOLDSMITHS UNIVERSITY OF LONDON

2010

The "Photostroller" is an interactive product developed for retirement a community in order to enhance the daily lives of its residents. The product displays a looped sequence of images drawn from the Internet, related and random, reminiscent of an electronic daydream. The flow can be influenced to stay close to a selected category of images, or allowed to drift away to more tenuously related subjects.

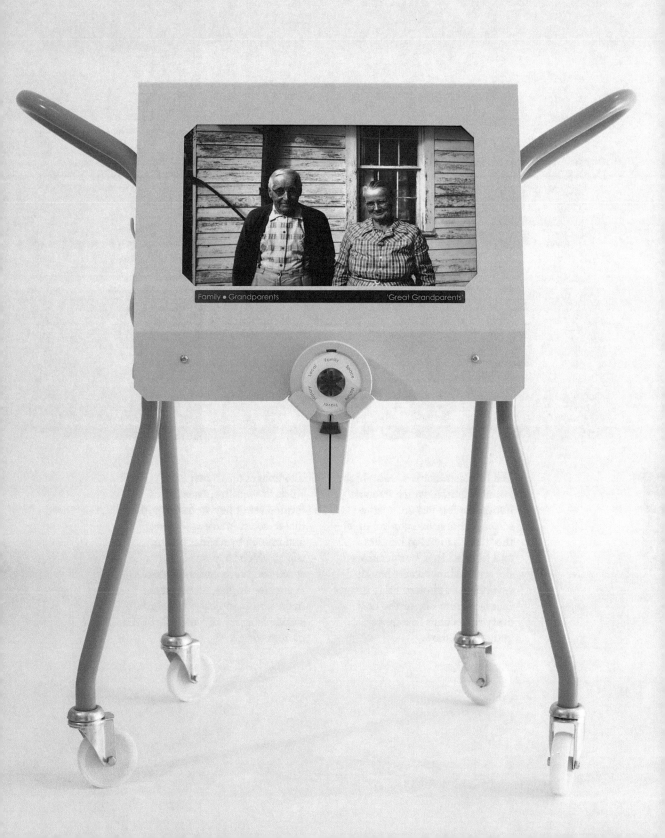

Family ● Grandparents   'Great Grandparents'

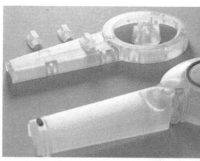

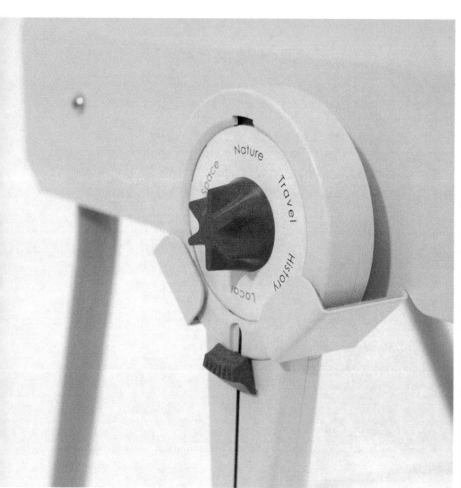

P

The Photostroller is a waist-high, portable unit, in which its most noticeable features are a large screen, a removable control unit—the "tuner", rounded handles and wheeled legs. In operation, the screen shows a continuous succession of photographic images that fade from one to the next every six seconds in a never-ending slideshow.

The images are drawn from Flickr™ websites, accessed via a limited set of keywords using the dial to select image categories, and refined by a slider on the tuner, which shapes the search space and holds current images. A smaller display, set under the main screen, indicates the current search category and the title of the current image.

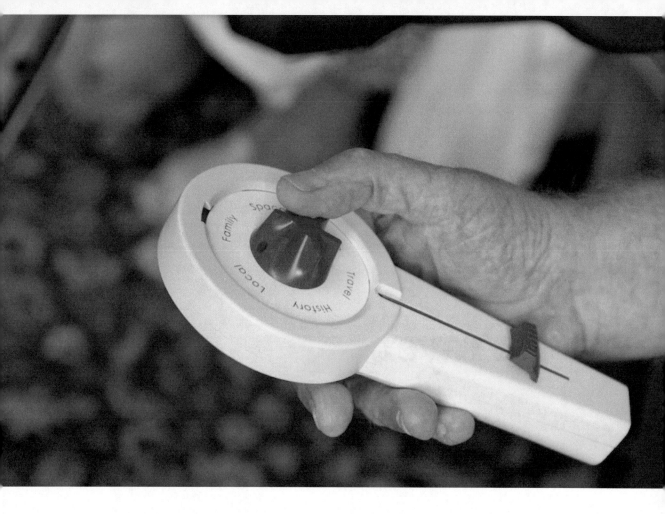

Interaction behaviors among older adults were carefully studied prior to finalizing the design. The type of interactions informed the design of a wireless controller to enable the residents to tune the type of photographs displayed in the slideshow. Emphasis was given to design a controller with relative ease of use.

Observations during its installation in the care home confirm the residents' engagement with the Photostroller over an extended period of time. Residents have taken responsibility for turning it on and explaining it to others, and its mobility has made it possible to use flexibly within the community.

MORE INFORMATION:

Gaver, W., Boucher, A., Bowers, J., Blythe, M., Jarvis, N., Cameron, D., Kerridge, T., Wilkie, A., Phillips, R., & Wright, P. (2011). The Photostroller: Supporting Diverse Care Home Residents in Engaging with the World. Proceedings from SIGCHI'11: Conference on Human Factors in Computing Systems. Vancouver, BC, Canada.

Content in this section provided by William Gaver.

Prayer

Companion

INTERACTION RESEARCH STUDIO,
GOLDSMITHS UNIVERSITY OF LONDON   2009

The "Prayer Companion" is an interactive device designed for a nuns community. Taking vows of poverty and enclosure, the nuns lead a simple life bound within the walls of the monastery, where the days are tightly structured around prayer.

The Prayer Companion is designed to provide a resource for prayer, with content from a wide range of media resources. In order to avoid media bias affecting prayers, the device delivers news headlines from around the globe.

P

P

The Prayer Companion is a wireless device. It functions when connected to the internet. Short sentences about current news or people's feelings are continuously fed to the device. The sentences scroll across the screen reporting up-to-date events and concerns, which are taken from a wide variety of global news sites as well as excerpts from social networking sites where people can write about their experiences and emotions. The intention is both to capture a sense of everyday concerns, and to alert the nuns to those who might need intervention yet don't ask for it -- but might Tweet about it.

The Prayer Companion have been implemented in the community since 2009, helping the nuns to keep their prayer pertinent.

MORE INFORMATION:

Gaver, W., Blythe, M., Boucher, A., Jarvis, N., Bowers, J., & Wright, P. (2010). *The Prayer Companion: Openness and Specificity, Materiality and Spirituality. Proceedings from SIGCHI'10: Conference on Human Factors in Computing Systems. Atlanta, Georgia, USA.*

http://www.huffingtonpost.com/2011/07/28/moma-talk-to-me_n_911841.html

Content in this section provided by William Gaver.

Conclusion

The book contains well-documented and illustrative recent examples of designed technologies for the aging population in the broad areas of design and computation—ranging from wearable devices, to mobile applications, to assistive robots. The presented work is a result of a research study in which industry, laboratories, and learning institutions were surveyed using a snowball sampling technique in order to idetify and document up-to-date projects on the topic. Even though snowball sampling was a suitable non-probability method for visualizing a network as well as recruiting and enlarging the database of projects[1], most of the contributors failed to recommend future acquaintances. Most contributors were identified through comprehensive searches

1. Visser, P. S., Krosnick, J. A., & Lavrakas, P. J. (2000). Survey research. In Handbook of Research Methods in Social and Personality Psychology (pp. 223-252). New York, NY: Cambridge University Press.
2. Groat, L. N., & Wang, D. (2002). Qualitative Research. In Architectural research methods. New York, NY: J. Wiley.

on university projects, journals, online blogs, to mention a few. The search led to a large sample, in which a vast number of projects did not meet the established criteria of having been designed within the past five years, while also making use of digital interactive technologies. Several contributors failed to submit sufficient documentation on their projects. As a result, 37 projects were selected based on the aforementioned criteria, merit, and creativity. The selected projects were analyzed, in which four main areas emerged: *social connections*, *independent self care*, *healthy home*, and *active lifestyle*. The projects were also coded[2] and further analyzed under two measures: type of design and technology features. The "type of design" included 46% being a *prototype design*, 30% being a *design in the marketplace*,

Figure 1: Type of design.

and 24% being a *concept design* (see Figure 1). Further analysis were conducted, in which 20 categories emerged for the "type of technology features", including *adaptive interfaces, broadband connections, data management, design research, embedded computing, graphic interface design, interaction design, internet connected, mobile applications, robotics, sensing technologies, social media, tangible interfaces, touch screen interfaces, universal design, vision interfaces, video communications, wearable devices, web applications*, and *wireless technologies* (see Figure 2).

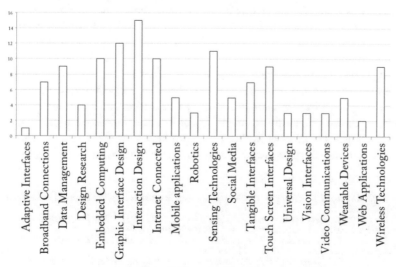

Figure 2: Type of technology.

Coded projects data show that *interaction design* was a present characteristic in 40.5% of all cases in this book. Even though it could be argued that all entries in this book may fall into the interaction design[3] characteristic, in this book the term refers to technologies that that have been designed from a system perspective, meaning that one or more products have been shaped to interact with one another and with the user. This evidence may suggest that not only is there a growing interest in de-centralizing product design as a system of interacting devices, but also visual interfaces are an important portal for the design of technologies. Also, *graphic interface design* was a present characteristic in 30.4% of all cases in this book.

In addition to *interaction design* and *graphic interface design*, *sensing technologies, embedded computing,* and *internet connected* were the most frequently present characteristics of all cases in this book. On the contrary, *adaptive interfaces, universal design, web applications, vision interfaces, video communications,* and *design research* were the least frequently present characteristics of all cases in this book. Of importance, *universal design*[4] was a present characteristic in 8.1% and design research was in 10.8% of all cases in this book. It is relevant to discuss these findings as universal design is a central design activity for ensuring the usability of a product for a wide range of population, considering older adult's varied abilities.

Overall, products presented in this book do not follow a common design practice, confirming the need for developing an approach toward designing technologies for older adults. The following details a framework proposition as a comprehensive and interdependent criteria.[5][6] First of all, *functionality* of the products should be simple to use addressing basic and specific older adults' needs. Second, products should exercise the use of *physical interfaces*[7]—hardware

3. *Saffer, D. (2010). Designing for Interaction: Creating Innovative Applications and Devices (2nd ed.). Berkeley, CA: New Riders.*
4. *The Center for Universal Design, North Carolina State University (1997). The Principles of Universal Design. From http://www.ncsu.edu/ncsu/design/cud/about_ud/udprinciplestext.htm*
5. *Rébola, C. B., & Jones, B. (2011). Sympathetic Devices: Communication Technologies for Inclusion. Physical and Occupational Therapy In Geriatrics, 29(1), 44-58.*
6. *Rebola, C. B., & Jones, B. (2013). Sympathetic Devices: Designing Technologies for Older Adults. Proceedings from SIGDOC'13. Greensboro, NC, USA.*

components to manipulate digital actions as they can afford more accessible interfaces for older adults. Third, beyond physicality of the interface, *contextualization* should be incorporated in the design. Contextualization refers to the relationship of situation, location, and space in which the technology is used. Fourth, there are a number of *design research methods*[8][9] that can facilitate the design of products, especially co-design or participatory methods[10]. The successful design of technologies arouses from involving older adults as experts in the design process.

New technologies should be designed in partnership with older adults[10] as individuals to jointly develop and test new products, systems, and services to address issues from mobility, to eating, staying fit, communication, or health—related conditions. Fifth, *universal design*[4] principles must be exercised in the design. Those principles have been developed

for almost 20 years in order to guide an effective design for all, in which older adults are a diverse population with varying abilities. Sixth, even though it is not a principle, *enjoyment* of the product should outweigh the effort of use. It is not just about fullfillment of real needs but, more importantly, emotional satisfactin in the product-user relationship.

The framework proposition is a simple step toward agreeing on best technology design practices for the aging population. It is imperative to generate innovative solutions to help older adults stay well, and live independently for longer and fulfilling lives. Design can make a difference on what works, who uses it, and why a technology lives with older adults. This book presented a compilation of designed technologies for older adults, serving as an example to promote the effective applications-design of technology-enabled systems and services to improve the quality of life of older adults.

7. O'Sullivan, D. and Igoe, T. (2004). *Physical Computing: Sensing and Controlling the Physical World with Computers.* Boston, MA: Thomson.

8. Martin, B. and Hanington, B. M. (2012). *Universal Methods of Design: 100 Ways to Research Complex Problems, Develop Innovative Ideas, and Design Effective Solutions.* Beverly, MA: Rockport Publishers.

9. Kumar, V. (2013). *101 Design Methods: A Structured Approach for Driving Innovation in your Organization.* Hoboken, NJ: John Wiley & Sons, Inc.

10. Sanders, E. B., Brandt, E. and Binder, T. (2010). *A Framework for Organizing the Tools and Techniques of Participatory Design.* Proceedings from PDC'10: The 11th Biennial Participatory Design Conference. Sydney, Australia.

Contributors

Ben Arent
Ronald Baecker
Maarten Bodlaender
Fred Bould
Robert Burke
Kelly E. Caine
Ya-Fang Cheng
Szu-Yang Cho
Celine Chhoa
Silvana Cieslik
Ben Davies
John Davis

Neil Dawson
Alexandra Deschamps-Sonsino
Ellen Yi Luen Do
Stacey Force
Larry Freil
William Gaver
Mudit Gupta
James Hallam
Tom Harries
Vincent Jeanne
Mikael Johansen
Masayoshi Kanoh

Emily Keen
Peter Klein
Hyungsin Kim
Dew-Anne Langcaon
Simon Levi
Siqi Liu
Paul MacKinnon
Karl Maybach
Steve Meinster
Joanna Montgomery
Natalie Montgomery
Mark Morgan
Lee Murray
Arun Nagargoje
Nitya Noronha
Yukio Oida
Lizzie Ostrom
Angela Paris

Graham Pullin
Eva Rielland
David Rose
Lauren Schmidt
Ned Semonite
Taro Shimizu
Robbert Smit
Carrie Smith
Philip Smith
Tomas Sokoler Medplast
Michael Spear
Hsien-Hui Tang
Joules Toulemonde
Luigi Trabucco
Egle Ugintaite
Michail Vanis
Hazel White
Chan Po Yee

Companies,
Laboratories
and
Institutions

*Aware Home Research Initiative,
Georgia Institute of Technology*

Big Bang Inc.

Bould Design

Centagon

Chang Gung University

Chukyo University

Clarity, Plantronics

*Ethical Technology in the Homes
of Seniors, Indiana University-
Bloomington*

*D-Matters Studio Lab, Georgia
Institute of Technology*

Georgia Institute of Technology

Good Night Lamp Ltd

Honeywell, HomMed

Hong Kong Polytechnic University

iHealthHome

Interaction Research Studio, Goldsmiths University of London

IT Unversity of Copenhagen

Lively

Design Information and Thinking Lab, National Taiwan University

National University of Taiwan of Science and Technology

Ode

Philips

School of Industial Design, Georgia Institute of Technology

Sonamba

Technologies for Aging Gracefully (TAG) Lab, University of Toronto

Duncan of Jordanstone College of Art & Design, University of Dundee

User Interface Design GmbH

VGo Communications Inc.

Vitality

YoooM, PresenceDisplays

Author

Dr. Claudia B. Rebola is an Associate Professor in the Industrial Design Department at Rhode Island School of Design. Dr. Rebola was previously an Assistant Professor and Director of Graduate Programs in the School Industrial Design and head of the D-Matters Studio Lab at the Georgia Institute of Technology. She is co-founder of the Design and Technologies for Healthy Aging (DATHA) initiative housed at the Center for Assistive Technologies and Environmental Access (CATEA). Her work brings together design, science, and technology to experiment, design, and prototype innovative interactive products in the realm of communication and social interactions. Her specific interests are in application areas tailored to design for aging with an emphasis on humanizing technology, empowering users, and celebrating the value of simplicity and tangibility in user-product interactions. As a native of Argentina, she received her undergraduate degree in Industrial Design from the Universidad Nacional de Cordoba. She also holds a M.S. in Industrial Design and a Ph.D. on Information Design from North Carolina State University. Dr. Rebola has been the recipient of numerous honors and awards, among them the Fulbright-Hays Grant, Phi Kappa Phi, Tau Sigma Delta, and an Argentinean presidential recognition for outstanding academic achievements in the design discipline.

Printed in the United States
by Baker & Taylor Publisher Services